T0291713

# Sacrifice Zones

# Sacrifice Zones

## The Front Lines of Toxic Chemical Exposure in the United States

Steve Lerner

*foreword by Phil Brown*

The MIT Press
Cambridge, Massachusetts
London, England

First MIT Press paperback edition, 2012

This book was set in Sabon by Graphic Composition, Inc.

Library of Congress Cataloging-in-Publication Data

Lerner, Steve.
Sacrifice zones: the front lines of toxic chemical exposure in the United States / Steve Lerner.
    p.   cm.
Includes bibliographical references and index.
ISBN 978-0-262-01440-3 (hc. : alk. paper)—978-0-262-51817-8 (pb. : alk. paper)
1. Environmental toxicology—United States—Case studies. 2. Chemical spills—Health aspects—United States—Case studies. 3. Hazardous substances—Health aspects—United States—Case studies. 4. Hazardous waste sites—United States—Case studies. 5. Pollution—United States—Case studies. I. Title.
RA1226.L47   2010
363.738'420973—dc22

                                                                    2009051289

For Sonya

# Contents

# Foreword

In a stratified society where race and class separate individuals, neighbor-hoods, and whole communities, people of color always know that their homes, roads, schools, utilities, parks and ball fields, sanitation services, and police protection are inferior to whiter and wealthier places. So it has not been a big leap for them to see that similar inequality exists in environmental contamination. For about as long as there has been a toxic activist movement—dating largely from Love Canal in the late 1970s—an environmental justice movement has existed as well, even if all the participants do not yet use and grasp the term *environmental justice*. The heroics of black, Latino, Native American, Inuit, and white working-class people in local toxic activist groups have made an indelible mark on everything we know about environmental contamination, health effects of toxics, remediation strategies, environmental social movements, regu-latory policies, and public awareness.

Steve Lerner recently showed us this harsh reality in his wonderful book *Diamond: The Struggle for Environmental Justice in Louisiana's Chemical Corridor*, about a black community in Louisiana striving for a corporate buyout of their homes in the polluting shadow of a major Shell refinery. With his journalistic eye and environmentalist ethics, Steve took us into the homes and hearts of the common people fighting an un-common fight. *Diamond* followed in the tradition of case studies about such toxic struggles that were nourished by Kai Erikson's *Everything in Its Path: The Destruction of Community in the Buffalo Creek Flood* and then continued with Adeline Levine's *Love Canal: Science, Politics, and People*, Michael Edelstein's *Contaminated Communities: The Social and Psychological Impacts of Residential Toxic Exposure*, Steve Kroll-Smith and Stephen R. Couch's *The Real Disaster Is above Ground: A*

*Mine Fire and Social Conflict*, Michael Reich's *Toxic Politics: Responding to Chemical Disasters*, Lee Clarke's *Acceptable Risk: Making Decisions in a Toxic Environment*, Steven Picou's *Social Disruption and Psychological Stress in an Alaskan Fishing Community: The Impact of the Exxon Valdez Oil Spill*, and my own *No Safe Place: Toxic Waste, Leukemia, and Community Action*. These studies gave us in-depth understanding of ordinary people as they discover the contamination and resulting health effects that typically plague them; the mobilization to get recognition, cleanup, and compensation; the resistance of the corporate and military polluters; the failures of state and federal regulatory agencies; and the development of key grassroots organizers and activist groups to coordinate efforts across those phenomena.

Having done that in *Diamond*, Steve Lerner wanted to tell a larger number of stories about other fenceline communities where the massive contamination was literally in their yard. This led him to continue his investigation by compiling a dozen case studies of communities across the United States, selected to cover the gamut of racial, ethnic, and class composition, as well as the variety of types of contamination and the types of success. Lerner guides us to understand the commonalities across these twelve places—the features of the in-depth understanding I point to in the previous paragraph.

This is a field I have been studying, writing about, and acting in since the mid-1980s, and I feel a fresh voice in Lerner's work. In reading these case studies, I was glad to meet new communities and to revisit with other activist communities I have been involved with. I know what these voices sound like and how hard it is to put them on paper. Lerner captures people's experience, suffering, realization, motivation for action, support for each other, and courage over long hauls that routinely last beyond a decade. His sensitive portrayal of these activists makes us feel as if we are in the midst of their situations—the flaring refineries and oozing hydrocarbons, the patronizing executives and obstinate public officials at open meetings, the piles of documents retrieved, the laypeople capturing air samples and digging test wells, the frustration of good-hearted community residents, the cries of parents burying their children.

If the parties responsible for pollution and cleanup lived in that midst, my guess is that they'd be quicker to prevent disaster and remediate any that occur. If they could muster the compassion and respect Lerner shows for people in contaminated communities, we'd hope they'd be less likely to point to explanations about individual responsibility, such as smoking and diet, and to grasp the widespread destruction of a deathly system of chemical production, distribution, and disposal.

Maybe some who need to will listen. Based on his case studies, Lerner offers many suggestions for change, with roles for every possible party: the federal government, the "big 10" major environmental organizations, grassroots activists, and firms with a notion of corporate responsibility. He suggests a variety of methods: a halt to siting heavily polluting facilities in already overburdened areas, protective zoning, tax incentives for green production, better regulatory enforcement overall in tandem with stricter environmental standards in areas of high contamination, more protective exposure guidelines, study and guidelines for the endless number of unregulated and unstudied toxics, and building political empowerment in communities where toxic hazards are most prevalent.

There is so much to do to prevent toxic trespass and environmental injustice. Thanks to Steve Lerner for helping move this along in *Sacrifice Zones*.

Phil Brown, Professor of Sociology and Environmental Studies, Brown University, author of *Toxic Exposures: Contested Illnesses and the Environmental Health Movement*

# Preface

Five years ago I wrote a book about a community in Louisiana sandwiched between a massive refinery and chemical plant. *Diamond: A Struggle for Environmental Justice in Louisiana's Chemical Corridor* (MIT Press, 2005) tells the story of what life was like for people in a segregated town twenty-five miles west of New Orleans who had the misfortune to live next door to two heavily polluting industrial plants. The book describes how Margie Richard, a local school teacher, organized a grassroots antipollution campaign that ultimately succeeded in forcing one of the world's largest oil companies, Shell Oil, into buying the homes of local residents so they could relocate out of harm's way.

After completing the book, it became clear to me that what had occurred in this tiny African American neighborhood, located on the banks of the Mississippi River, was in no way unique; there are in fact thousands of communities on the fenceline with heavy industry where low-income people of color (and some whites) continue to be disproportionately exposed to toxic chemicals.

To demonstrate the scale of the problem, I decided to write a follow-up book that provides a dozen case studies about what life is like on the wrong side of the tracks in contaminated communities across the nation, from New York to California and from Florida to Alaska. My goal is to inject into the public dialogue the voices of fenceline residents who live under these difficult circumstances, describe in detail what they experience, what they have in common, and how they are organizing to improve their quality of life and protect themselves.

Were most Americans to visit one of these sacrifice zones and stay for a few days, I am convinced they would come to the conclusion that residents who reside near heavy industry should not have to live under these intensely polluted conditions and that, as a nation, we can do better in protecting our citizens, of whatever race and income level, from these widespread environmental injustices.

# Acknowledgments

Many people helped make this book possible. Chief among them is my brother, Michael Lerner, president of Commonweal, who has supported my work for the past thirty years. For his wise counsel, his editorial suggestions, his patience, and his collaborative spirit, I am deeply grateful.

I also thank my colleagues past and present at Commonweal and at the Collaborative on Health and the Environment, including Elise Miller, Eleni Sotos, Sharyle Patton, Davis Baltz, and Charlotte Brody.

Special thanks are also due to John and Ann Doerr, Lucy Waletsky, and Marion Weber, who provided the financial support that allowed me to do the research for this book

I have benefited greatly from the help provided by Denny Larson and Ruth Breech at Global Community Monitor, who introduced me to a number of the grassroots leaders profiled in these pages, and to Wilma Subra who shared her knowledge about a number of these fenceline struggles.

This book would not have been possible without the cooperation and generous help provided to me by the many residents of the sacrifice zones I profile in the following pages. Despite many hardships and a busy schedule, they took me into their homes and spent hours explaining to me what life was like on the fenceline with heavy industry and military bases. I will always admire their courage and persistence in their struggle for environmental justice.

Finally, I am indebted to Clay Morgan at the MIT Press who made it possible for this and my previous books to see the light of day. I would also like to thank Sandra Minkkinen for her help editing the manuscript.

# Introduction

"I just got mad. I couldn't breathe in my own house. The fumes smelled like the gas stove was on but not lit. I'd run downstairs to check the stove to see if I had left a burner on. It smelled like lighter fluid," recalls Ruth Reed, a retired elementary school teacher and resident of Ocala, Florida, who lives next door to a charcoal plant.

The reason Reed's home smelled like an unlit gas stove was not hard to detect. Looking out her window, she could see a plume of smoke rising from the Royal Oak charcoal factory's chimney that was shrouding the neighborhood in soot. There was nothing wrong with Reed's stove; it was the whole neighborhood that was engulfed in a cloud of smoke and lighter fluid fumes. "I'd wet a washcloth and put it over my face. I couldn't open the windows. I just got sick and tired of it," says Reed, explaining why she organized her neighbors to protest the pollution.

Reed is not alone in saying, "Enough is enough." Hundreds of fence-line residents living adjacent to heavily polluting plants reach this push-back point. They get fed up when their homes and bodies are invaded by highly toxic releases from nearby facilities. After breathing in large volumes of polluted air or swallowing countless gallons of poisoned water, some of them finally take action: organize their neighbors, speak up at public meetings, research the health effects of their chemical exposure, invite the media to report on the contamination problems, and kick up a ruckus.

The chapters that follow recount environmental justice struggles in a dozen pollution-afflicted fenceline communities and celebrate the courage and perseverance of grassroots leaders who protest the contamination that permeates their neighborhoods. Most of these residents are school teachers, veterans, health care professionals, or retired people. They do not have prior experience as activists and do not see themselves, at least initially, as environmentalists. For the most part, when the local pollution problem first comes to their attention, they have been living private lives and raising their families. But when the fumes become too intense, when they find their family and friends falling ill from pollution-induced disease, they shed their quiet ways and organize a protest.

What follows are the stories of some of the leaders of this grassroots environmental justice movement that is gathering momentum across the nation. In Pensacola, Florida, Margaret L. Williams became alarmed at the high rates of illness in her community located next to two Superfund hazardous waste sites and organized her neighbors to demand relocation.[1] In Port Arthur, Texas, Hilton Kelley, a retired Navy man and actor, returned to find dangerously polluted conditions at the public housing project where he grew up located across the fence from a huge Exxon-Mobil refinery. He rallied his neighbors to protest the pollution and demand that improved pollution control equipment be installed. And in Tallevast, Florida, Laura Ward looked out her window and saw technicians drilling a hole in her lawn to see if pollution from the neighboring Lockheed Martin weapons plant had contaminated the groundwater. Ward, along with Wanda Washington, brought together members of their community to press for a cleanup of groundwater and compensation for harm done to their health, quality of life, and property values.

### Frontlines of Toxic Chemical Exposure

What the residents profiled here have in common is that they all live on the frontlines of toxic chemical exposure in the United States. They inhabit low-income "sacrifice zones" where hundreds of thousands of residents are exposed to disproportionately elevated levels of hazardous chemicals. They reside in semi-industrial areas—largely populated by African Americans, Latinos, Native Americans, and low-income whites—where a dangerous and sometimes lethal brand of racial and economic discrimination persists.

The label *sacrifice zones* comes from "National Sacrifice Zones," an Orwellian term coined by government officials to designate areas dangerously contaminated as a result of the mining and processing of uranium into nuclear weapons. During the Cold War, when the Soviet Union and the United States were racing to build up their nuclear arsenals, large areas in both nations were contaminated with radioactivity. In the United States some of these catastrophically polluted places were fenced off and warning signs were posted; but others were not, and people continued to live in them and fall ill. Today hundreds of these national sacrifice zones are scattered across the United States, where the by-products of uranium mining operations, nuclear weapons production facilities, and atomic test sites have left behind irradiated landscapes unfit for human habitation. As early as 1988, some seventeen principal weapons production sites in twelve states were identified as needing to be fenced off or cleaned up, and another eighty locations in twenty-seven states were awaiting further action.[2]

Areas contaminated with radioactivity are not the only places "sacrificed" to the ravages of intense pollution. In the following pages, I make the case that the "sacrifice zones" designation should be expanded to include a broader array of fenceline communities or hot spots of chemical pollution where residents live immediately adjacent to heavily polluting industries or military bases. While government officials concede that the production of nuclear weapons regrettably caused a small number of citizens to make health and economic sacrifices on the altar of national security, they ignore a much larger host of low-income and minority Americans whose health is sacrificed as a result of chemical contamination.

A number of efforts have been made to find a label that succinctly describes these residential/industrial areas that experience severe contamination problems. In legislative hearings they have been called *environmental high-impact areas*, but this technical locution did not last and is almost never heard today. Environmental justice activists, who work to improve conditions in these blighted areas, tend to call these areas "sacrifice zones," "fenceline communities," or "hot spot of pollution." I have chosen to highlight the first of these descriptors, "sacrifice zones," in the title of this book because it dramatizes the fact that low-income and minority populations, living adjacent to heavy industry and military bases, are required to make disproportionate health and economic sacrifices that more affluent people can avoid. To my mind, this pattern of unequal exposures constitutes a form of environmental racism that is being played out on a large scale across the nation.

## Environmental Justice Awakening

For decades little was heard from these fenceline communities. People lived within them—drinking polluted water and breathing contaminated air—without seeing a way to do anything about it. A kind of fatalism about being trapped within these polluted precincts prevailed. Environmental conditions in these neighborhoods were often so bad that it was obvious (at least to those who lived there) that these conditions were making people ill, but that was just the way it was. Residents in these sacrifice zones where the factories were concentrated knew that more affluent whites did not have to endure the kind of heavy pollution that rained down on their side of town, but few of them could afford to protect themselves by moving.

Then in the 1980s, something changed. The sense of fatalism that many residents in sacrifice zones experienced began to lift. The awakening began in 1979 when residents in an African American suburb of Houston filed a lawsuit against Browning-Ferris Industries for environmental discrimination in siting a municipal solid waste landfill in their community. In the lawsuit, *Bean v. Southwestern Waste Management*, the plaintiffs noted that six of Houston's eight solid waste landfills were located in African American neighborhoods, despite the fact that African American residents in the city comprised only 28 percent of Houston's population.[3] Subsequently, in 1982, some four hundred protesters were arrested in Warren County, North Carolina, where local officials were determined to bury thirty-two thousand cubic yards of soil contaminated with highly toxic polychlorinated biphenyls (PCBs) in a community that was 82 percent African Americans. In 1983 a General Accounting Office study found "a strong relationship" between the location of off-site hazardous waste landfills and the race and socioeconomic status of surrounding communities in southern states. The majority of the population was African American in areas where three of four landfills were sited.[4] Then in 1987 the United Church of Christ published "Toxic Waste and Race in the United States: A National Report on the Racial and Socioeconomic Characteristics of Communities with Hazardous Waste Sites." The study "showed an unmistakable statistical and spatial correspondence to minority populations" in the siting of hazardous waste dumps nationwide.[5]

Since these pioneering environmental justice studies were published, numerous new studies have been conducted that demonstrate the unequal chemical exposure burden suffered by many low-income and heavily minority communities. After reviewing sixteen environmental justice studies published between 1971 and 1992, two academic researchers concluded that there exists a "clear and unequivocal class and racial bias in the distribution of environmental hazards."[6]

One widely publicized study shows that African Americans are "79 percent more likely than whites to live in neighborhoods where industrial air pollution is suspected of posing the greatest health danger." These calculations were made by David Pace, a reporter for the Associated Press, who used Environmental Protection Agency (EPA) emissions data and U.S. Census figures. According to the data he collated in twelve states "Hispanics are twice as likely as non-Hispanics to live in neighborhoods with the highest [air pollution] risk scores." Carol Browner, former head of the EPA, acknowledged these sad facts: "Poor communities, frequently communities of color but not exclusively, suffer disproportionately." Furthermore, people within these sacrifice zones are poorer than average Americans and are 20 percent more likely to be unemployed.[7]

State-specific environmental justice studies have detected a similar pattern of unequal exposure rates. One study of 368 communities in Massachusetts found that "low-income communities face a cumulative exposure rate to environmentally hazardous facilities and sites that is 3.13 to 4.04 times greater than that of all other communities [in the state]. . . . In addition, high-minority communities face a cumulative exposure rate to environmentally hazardous facilities and sites that is nine times greater than that for low-minority communities." Low-income and minority communities in Massachusetts are also located closer than other populations to heavily polluting industrial facilities, hazardous waste sites, landfills and transfer stations, incinerators, and power plants, the authors observe. "Clearly, not all communities in Massachusetts are polluted equally—lower income communities and communities of people of color are disproportionately impacted," they conclude.[8]

Fines levied on polluting industries are also unequally apportioned. An examination of eleven hundred Superfund sites reveals that "the average fine imposed on polluters in white areas was 506 percent higher than the average fine imposed in minority communities." Cleanup efforts in minority communities also took longer despite the fact that they were less intensive than those in white areas.[9]

These hard facts provided ammunition for the rapidly coalescing environmental justice movement. As word of this research began to circulate, many residents in polluted zones became aware that their experience of severe exposures to toxins was part of a nationwide pattern of racial and economic environmental injustice. Today some two hundred grassroots activist groups work on contamination problems in communities with a high concentration of people of color.[10]

## Dangerous Zoning Decisions

Sacrifice zones are the result of many deeply rooted inequities in our society. One of these inequities takes the form of unwise (or biased) land use decisions dictated by local or state officials intent on attracting big industries to their town, county, or state in an effort to create jobs and raise tax revenues. When decisions are made about where to locate heavily polluting industries, they often end up sited in low-income communities of color where people are so busy trying to survive that they have little time to protest the building of a plant next door. Those who make the land use decisions that govern sacrifice zones typically designate these areas as residential/industrial areas, a particularly pernicious type of zoning ordinance. In these areas, industrial facilities and residential homes are built side by side, and few localities have adequate buffer zone regulations to provide breathing room between heavy industry and residential areas.

The health impact of this patently unwise zoning formula is predictable: residents along the fenceline with heavy industry often experience elevated rates of respiratory disease, cancer, reproductive disorders, birth defects, learning disabilities, psychiatric disorders, eye problems, headaches, nosebleeds, skin rashes, and early death. In effect, the health of these Americans is sacrificed, or, more precisely, their health is not protected to the same degree as citizens who can afford to live in exclusively residential neighborhoods.

Behind the cold statistics about unequal exposure rates are real people, and the policies that permit disproportionately high chemical exposures in sacrifice zones have flesh-and-blood consequences. This book gathers together hundreds of eyewitness accounts about what life is like in these fenceline communities and the sacrifices the residents have been forced to make. The research for this work was reportorial in nature. Over the course of two years, I traveled to a dozen sacrifice zones. There were hundreds of these heavily polluted, residential/industrial areas to choose among. I decided to focus on communities where the environmental justice struggle was already mature and grassroots groups had already organized, begun to protest environmental conditions, lobbied for reduced emissions or relocation, and garnered some local media attention. In all, I interviewed hundreds of residents in these neighborhoods, convinced that fenceline residents were the real experts and were best positioned to describe the conditions near their homes.

One of those I met was Robert Alvarado Sr., who lives in San Antonio, Texas, next door to Kelly Air Force Base. Alvarado and his family reside on top of a plume of groundwater contaminated when military personnel dumped solvents used to clean airplane parts into a hole in the ground. One of the solvents they dumped was trichloroethylene (TCE), a chemical listed as a probable carcinogen and linked to liver and kidney cancer, as well as birth defects. Alvarado, probably not coincidentally, suffers from severe liver and kidney disease.

"I'm angry. They shortened my life. I can't prove it, but I'm more than suspicious," says Alvarado whose feet are often so swollen that he cannot wear socks. "They cut my work-life short." As a result of living next to an intense source of pollution, his wife, Guadalupe, instead of retiring, had to go back to work at a dry cleaners despite the fact that she is also sick.

## Common Problems in Sacrifice Zones

In his epic novel *Anna Karenina*, Leo Tolstoy observes that all happy families are happy in much the same way, while unhappy families are unhappy in their own particular way. In a parallel fashion, communities in which environmental quality is good have much in common (e.g., they tend to be relatively affluent, majority white, and not located near heavy industry), while the contaminated ones are each distressed in their own special ways.

Conditions in every sacrifice zone are unique. The type of industry or military base next door that emits large volumes of contaminants varies from one place to another, as do the specific toxic chemicals that waft over the fenceline into the residential neighborhoods. The geography of sacrifice zones also ranges widely, from inner-city neighborhoods to remote rural locations. Yet despite these differences, commonalities run as recurrent themes through the twelve case histories presented in this book.

To begin, it is frequently the case that fenceline residents experience a moment of rude awakening—an awful surprise—when they learn that they have been exposed to elevated levels of toxic chemicals for years and sometimes decades. This wakeup call often comes when residents observe officials drilling test wells or installing air monitors dressed in protective hazmat gear that makes them look as if they are wearing space suits. These extraordinary precautions make residents ask themselves: "Well, if they need this protection, what about me and my family?"

Once fenceline residents begin asking questions about the extent of the contamination, they frequently report that they get the runaround from officials and are rarely given straight answers or comprehensive information. They often later learn that both government and corporate personnel withheld the bad news about the extent of the contamination out of concern that it might create a panic. This deceitful, paternalistic behavior makes it impossible for local residents to make timely and informed choices about whether to move immediately (if they can afford to), stop drinking well water, keep the windows closed, send their children to live with relatives in other neighborhoods, prohibit their children from playing outdoors, avoid gardening or eating home-grown vegetables, or take other protective actions.

When residents learn that they are not getting the full story from government officials and corporate public relations representatives, they are thrown back on their own resources. It is at this point that the future leaders of grassroots groups tend to step forward and begin to do their own research. Their homes soon become repositories for stacks of abstruse technical reports, newspaper clippings, and reprints from medical journals. As they educate themselves about the contamination they face, they become amateur toxicologists and epidemiologists, as well as fluent in regulatory jargon. One resident told me, "I had to learn to talk in parts per million and parts per billion."

As they dig deeper into how their neighborhood came to be designated as a place where heavily polluting industries were sited, residents learn that the boundaries of industrial sacrifice zones are not arbitrary but instead are often shaped by racism. Most sacrifice zones are predominantly inhabited by African American, Latino, or Native American populations, although there are plenty of others where low-income whites live. Many of these neighborhoods were built in an era when segregation was still enforced. During this period, people of color frequently lived in designated areas that had undesirable characteristics: a location near a heavy industry, the railroad tracks, in a flood or earthquake zone, on the steep side of a hill, or at the edge of town near the county dump. Race zoning laws have since been struck down by the courts, and now everyone can, at least theoretically, live wherever they can afford to buy a home. But in the real world, the race lines left by segregation policies are still visible in many places as people of color pass their homes down to their offspring over the generations, and new residents are steered to racially homogeneous neighborhoods by real estate agents.

While the era of legally sanctioned racial segregation is past, a new form of not-so-subtle racism is occurring in which many low-income, heavily minority communities are designated as the unofficial dumping grounds for what is known among land use planners and real estate developers as locally unwanted land uses (LULUs). These LULUs are hard to miss for those willing to look. They include a wide variety of high-emission industrial plants and public utilities, including incinerators, hazardous waste dumps, refineries, gasoline tank farms, plastic plants, steel mills, pesticide plants, cement kilns, sewage treatment plants, rubber factories, asphalt batching plants, large-scale pig and cattle feedlots, agricultural areas heavily sprayed with pesticides, tanneries, machine shops, auto-crushing-and-shredding operations, and a host of other nasty facilities.

## Getting Organized

At a certain moment in the evolution of each grassroots environmental justice struggle, residents determine to get organized. This decision is usually galvanized by an event such as a cluster of pollution-induced illnesses or the release of a report or newspaper article revealing the extent of the contamination. Many of these groups come together when fenceline residents begin to ask themselves if the toxic releases they are just learning about might have caused the high rates of disease in their neighborhood.

The organization of these grassroots antipollution campaigns is not always without friction. In some cases, the affected community becomes divided over whether to make a public protest about contamination problems. Some residents worry that if their toxics problem becomes widely known, their property values will plummet, while others contend that they will not get any relief without speaking up. Communities are also sometimes divided as to whether to seek relocation, and renters and home owners often have different concerns when the possibility of relocation is being considered. Officials at industries accused of causing the pollution problem sometimes exploit these differing views in a divide-and-conquer strategy designed to splinter and outlast a local environmental justice campaign. In order to deal with corporate maneuvering and regulatory inaction, strong leadership is needed to keep residents unified, engaged, proactive, and inclusive in their demands so that the concerns of all factions are included.

Frequently, local antipollution group leaders come to realize that their community has neither the political clout nor resources to mount an effective campaign without outside support. At this point they begin to search for allies, only to find that there are a limited number of places they can turn to for help. At a national level, environmental justice experts such as Robert Bullard, who heads the Environmental Justice Resource Center at Clark Atlanta University, are often consulted.

Some of the other activists who provide help to grassroots leaders are Peggy M. Shepard at West Harlem Environmental Action, Lois M. Gibbs at the Center for Health, Environment, and Justice, Beverly Wright at Xavier University, Monique Hardin at Advocates for Environmental Rights, Bradley Angel at Greenaction, Anne Rolfes at the Louisiana Bucket Brigade, and Eric V. Schaeffer, currently director of the Environmental Integrity Project and former EPA official who quit his job in a protest over the inadequate enforcement of the Clean Air Act under the administration of President George W. Bush.

Local environmental justice activists also find they can get practical help learning how to carry out their own air monitoring from Denny Larson and Ruth Breech at Global Community Monitor; or help with regulatory and toxicology issues from Willma Subra, a chemist in Louisiana. These itinerant organizers teach fenceline residents how to keep odor and symptom logs, connect them with reliable government information that reports the volume of toxic chemicals being released by neighboring facilities, provide them with information about how to navigate the regulatory bureaucracy, and outline the potential health effects that various chemical exposures can cause. These national and regional environmental justice experts have helped many grassroots groups with technical, legal, funding, research, and strategy issues.

Even with help from outside experts and organizers, the learning curve required of local grassroots fenceline leaders is steep. Not only do they have to become knowledgeable about how to decipher abstruse regulatory reports and become conversant about toxic chemicals and their health impacts, they also must learn to be public speakers, attract media coverage, and give interviews in a way that makes a number of salient points in a quotable and memorable manner. Mastering the art of reeling off talking points backed up with hard data does not happen overnight, and many local activists only gradually become comfortable on the public stage. Learning to work the media, however, is often crucial to grassroots environmental justice struggles because corporations are highly sensitive to bad press and are sometimes willing to make concessions in order to avoid it. Similarly, the power of newspaper, radio, Internet, and television coverage to focus attention on a fenceline contamination problem is often one of the strongest assets that activists have available to leverage regulatory action that will bring relief to their community.

Taking on the big industry next door requires courage. What individual or small fenceline group really wants to risk the personal consequences of going up against the power of Lockheed Martin, ExxonMobil, or the U.S. military by accusing them of poisoning people? From the outset, fenceline activists are keenly aware of the monumental power imbalance between industries that emit the pollutants and the residents who breathe them. On one side of the fence is a group of residents who meet around a kitchen table or in a small church, talk about the odors coming from the plant, and complain to each other that they and their children are sick from the fumes. They make homemade signs to protest the pollution, plant crosses in the ground outside the homes of those they believe died from the contamination, and hold bake sales to try to raise the money to hire a lawyer.

On the other side of the fence stand companies with vastly greater resources at their disposal to fight the antipollution campaigns mounted by their residential neighbors. These companies have deep pockets, legions of lawyers, and public relations and community relations specialists to get out their message. They hire lobbyists in the state capital and in Washington to see that their interests are protected, and they have the money to fund charitable works in the community as well as the power to hire and fire local residents who work in their plants. The result of this power imbalance is not hard to guess: residents on this uneven playing field face an uphill struggle to force companies to reduce their emissions or provide funds for relocation.

After documenting a local pollution problem, one of the most effective initiatives that a grassroots group can launch is a door-to-door health survey. Many fenceline groups start by looking through cancer and birth defects databases for evidence of pollution-induced disease in their neighborhood. But these statistics are frequently not gathered in such a way that makes it possible to tell if the people next door to an industry are being affected. Often fenceline groups must do their own door-to-door health surveys to accumulate evidence of a health problem before they can prompt county, state, or federal health agencies to do a more formal inquiry. As we shall see, these informal resident-run health surveys can receive media coverage that leads to regulatory actions or further research.

Proving that a specific industrial chemical release led to a specific illness is notoriously difficult and even harder to prove in court. Even in cases where common sense makes it apparent that living immediately adjacent to a refinery has resulted in increased rates of asthma, respiratory disease, and other ailments, corporate lawyers are often successful in arguing that residents are sick because they are poor, lack access to medical care, eat poorly, smoke, or take drugs. This strategy of blaming the victim has served industry well over the years and can blunt legal challenges to their release of large amounts of toxic chemicals near residential areas. Almost all of the grassroots groups described in this book come up against the difficulty of proving that they have been harmed even when they have evidence of elevated levels of toxic chemicals in their air, water, or soil.

Finding lawyers who will take fenceline cases has proved difficult in many sacrifice zones. From the attorneys' perspective, proving health damages is a costly long shot that requires extensive research. As a result, in many instances, lawyers for fenceline residents try to prove that their clients' quality of life has been impaired or that their property values have been devalued as a result of the nuisance caused by the factory next door. But this fails to address the core issue in sacrifice zones: that people are being made sick and in some cases are dying as a result of being exposed to toxic releases from nearby facilities.

Given the difficulty of winning legal or regulatory relief for their pollution problems, environmental justice activists face stiff odds in winning their campaigns. Grassroots leaders often work without compensation for years, and sometimes decades, in an effort to protect the health of their community. At considerable risk to their safety, job security, and standing in their community, they organize meetings in churches and schools, and they speak to reporters. Their goal is rarely to shutter the industry causing the pollution. Instead, they work to improve enforcement of existing regulations, reduce emissions, and convince company owners to install state-of-the-art pollution control technologies. Only in extreme cases do they lobby for the shutdown of a noxious plant or the relocation of residents to homes in a less polluted area.

**Industry's Dilemma**

A description of sacrifice zones today would not be complete without acknowledging the complicated choices faced by the owners and managers of heavy industry, many of whom, despite easy caricature and vilification, want to do a good job producing a product—be it oil, coal, plastics, cement, or a place to get rid of wastes—while at the same time protecting public health. They invest tens of millions of dollars in pollution control equipment and have whole divisions within their firm devoted to compliance with environmental regulations.

These firms encounter real challenges in their efforts to operate in a manner that is both safe and profitable, and they face price competition at home and abroad to make their product cheaply and efficiently. Officials at American heavy industrial facilities argue that they are already at a competitive disadvantage because they must pay for costly pollution control equipment that foreign competitors are not required to install. They also have limited options about where they can build.

Take the case of a U.S.-based petrochemical company that wants to build a refinery near the Gulf Coast in Texas or Louisiana in order to refine millions of gallons of oil. These companies cannot just build anywhere. If they site their facilities in the middle of a desert, far from residential areas, they would have to pay huge sums to transport the oil overland instead of having it delivered cheaply and efficiently by super-tankers. And where is a stretch of Gulf Coast that is totally uninhabited? There isn't one. Thus, wherever they choose to build, they will discomfit some community.

Nevertheless, the regularity with which high-emission companies choose to build in low-income and heavily minority communities should be noted. This is no coincidence. These communities are targeted because there is less high-end development than in other areas, land is cheap in these neighborhoods, and residents are unlikely to offer organized resistance to the siting of their facility. Some industry officials have gone so far as to publicly describe this strategy: "A 1984 report by Cerrell Associates for the California Waste Management Board, for instance, openly recommended that polluting industries and the state locate hazardous waste facilities in 'lower socio-economic neighborhoods' because those communities had a much lower likelihood of offering political opposition."[11]

Not only do heavily polluting industries target low-income communities as sites for their facilities, once they buy land and set up shop, they often do little to accommodate the legitimate health concerns of their neighbors. Yes, they invest tens of millions of dollars in pollution control equipment. But they do not build their plants in such a way that no neighbors are hurt by their emissions. A regulatory code designed to protect the health of all Americans would require that these companies either reduce emissions to safe levels or buy up a buffer zone of land to keep residents at a safe distance from their emissions. Here, not surprisingly, industrialists point out that they are merely operating under an existing regulatory regime and that they are clearly meeting regulatory standards because their permits to operate have not been revoked by government officials.

## Hidden in Plain Sight

The fate of residents in sacrifice zones around the nation is not being debated in the corridors of power in Washington, D.C. This is unfortunate but not particularly surprising. David Halberstam, the late Pulitzer Prize–winning reporter, observed, "The farther one travels from centers of power the closer one gets to the truth." This is certainly true of sacrifice zones. To learn about what life is like in these fenceline communities requires traveling off the beaten track and venturing beyond the centers of affluence and power. Sacrifice zones are not garden spots and few people travel to them as destinations of choice. As a result, many of them remain essentially hidden from the view of most Americans.

The dozen case histories of fenceline communities in this book are designed to do the legwork for interested readers and bring them detailed accounts of how the grassroots environmental justice movement is gradually spreading as residents confronted with untenable pollution in one community hear about successful tactics used in another fenceline neighborhood. Within these sacrifice zones, the human cost of our rough-and-tumble, winner-take-all economic system is brutally visible. Here we can see the tragic consequences of our discriminatory zoning practices, our inconsistent standards about the health effects of toxic chemicals, and our gap-ridden regulatory system. Here we find Americans who cannot afford to live in a neighborhood where the air and water are clean and who are stuck instead in dangerously sited houses where they are literally choking on the exhaust of our industrial system.

Is this the America we want? If we are to begin to create a more equitable and sustainable culture in these United States, it is in communities such as these that we should begin the work and measure the outcome of our efforts.

# I

## Partial Victories

# 1

## Ocala, Florida: Community Blanketed by "Black Snow" from Neighboring Charcoal Factory

"You pray for it to rain," Ruth Reed told a reporter as she sat in her home in northwest Ocala, Florida, just a block from the Royal Oak furnace that baked scrap wood into charcoal briquettes used in backyard barbecues.[1] Reed and her neighbors prayed for rain not because of any drought but in hopes that a good downpour would clear the air of the fine soot and ash that fell like a black snow over their homes.

The soot covered everything in Bunche Heights, the largely African American, working-class neighborhood where Reed and her husband, Dr. Leroy Reed, had lived since the early 1960s, a decade before the Royal Oak factory opened in 1972. For the next thirty years, emissions from the factory spread a layer of white ash and black charcoal cinders over cars, laundry hanging up to dry, toys, wading pools, and the leaves of the trees. The soot discolored the siding on homes, prompting some residents to paint their houses a dark color to hide the stain.

Nowhere were residents safe from the charcoal plant's pollution. Soot infiltrated their homes, seeping through cracks beneath doors and into air-conditioning vents. "We can't open our windows unless it rains," explains Reed, sixty-six years old, who kept her house sealed up twenty-four hours a day.[2] City councilman Mike Amsden confirms that he has seen soot in many homes around the Royal Oak plant. As to its source, he says wryly: "It is not because they [the residents] are barbecuing indoors."[3]

### Staff Sergeant (Retired) Robert J. Brown

Reed was not the only one in the neighborhoods surrounding the plant affected and irritated by the pollution. A short drive from her house, in a neighborhood known as Happiness Homes, lived Staff Sergeant (retired)

Robert J. Brown, who served two tours in Vietnam with the U.S. Air Force. When Brown retired to Ocala, he thought his fighting days were over, but instead found himself on a different kind of frontline. After the Royal Oak charcoal plant was built and fired up, Brown became convinced that its soot threatened his health, that of his family, and everyone else in the neighborhood. The soot irritated his lungs and eyes so much he had to go to the hospital.

Having seen the world with the Air Force and served in Germany, Italy, Holland, Spain, and Vietnam, Brown knew that the clouds of pollution that descended on his home were not normal. He had hoped for a quiet retirement and enjoyed sitting on his front porch. His determination to take it easy was legendary: he went so far as to build a mechanical device that waved a wooden hand when he pulled a string. This allowed him to greet friends passing his home without exerting himself.

But Brown's quiet retirement was not to be. Every time Royal Oak ignited its furnace and ash and soot fell on his house, Brown climbed in his black pickup truck and drove down to talk with the plant managers. His complaints sometimes brought temporary relief, and the furnace would be turned off until the smoke cleared. On those occasions, he would return home triumphant and tell his family, "I think I've got them now." But a few hours later, the charcoal plant would restart, leaving him feeling deceived once again.

### Working the Media

Unwilling to give up, Brown next took his fight to the media and was featured in several local newspapers complaining about the soot. One early article includes a photo of his hand covered in soot from where he wiped the glass on his window. Another captures him with reddened eyes that were irritated from constant exposure to the soot. "It's like sandpaper in your eyes," he told a reporter. Brown went down to the VA hospital regularly to have his eyes irrigated and treated, but the emissions from the plant eventually affected his vision, he claimed.

Unable to convince plant managers to install pollution control equipment, Brown started gathering petitions protesting the plant's pollution and took them to city hall. He even collected samples of the soot in a small bottle to show city officials. His children remember that sometimes he would not let anyone move his truck for days so that it could gather soot undisturbed as evidence of the rate at which pollution accumulated. As part of his one-man campaign, he also wrote grievance letters to city, state, and federal officials, though without success.

All this time Brown's health was deteriorating. His coughing became so bad that he could no longer sit out on the porch because the pollution from Royal Oak would trigger a coughing fit. Other members of his family were also affected. Over the thirty-two years they lived near the charcoal plant, Brown watched as his wife, Jesena, and his foster daughter developed severe respiratory distress.

"The soot is killing us," Brown told his daughter, Kim Brown. "He just wanted to protect his family," she observes in an effort to explain her father's obsession with the plant. His children and grandchildren remember Brown proudly for his early efforts to close the plant. "Before I die that place will be shut down," Brown told his grandson, Otis Sumler IV. Brown believed the charcoal plant was built down the street from his family "because we live in an African American neighborhood," he told his daughter, Ann Brown. Although he was angry about the pollution, Brown did not want to sue anybody or get any money out of it, she recalls: he just wanted to get the plant shut down so he could breathe in his own home.

After years of breathing soot, Brown's cough intensified; he was taken to the hospital on a number of occasions and was eventually tethered to an oxygen tank at home and a mobile oxygen unit on the rare occasions when he went out. On May 13, 2007, he died of respiratory distress and congestive heart disease. He and his family believe that his illness was either the result of breathing in the pollution or at the very least was exacerbated by the soot. Many of those who attended his funeral praised Brown not so much for his military service to his country as his service to his community. "He was a neighborhood hero," says his daughter Kim Brown.

Brown's early battle against pollution from Royal Oak made sense to a lot of his neighbors, who were intimately familiar with the inky black smoke and tongues of fire that leapt from Royal Oaks smokestacks. Ruth Reed's husband, Leroy Reed, a professor of political science at Central Florida Community College, also made a number of trips to the city council from 1982 to 1996 to complain about pollution from the plant. "But the officials there would not bend," he recalls.

Following his example, other residents began speaking out. Betty Ann Smith described to a reporter the thick clouds of smoke as a part of daily life. About the air quality she said: "Sometimes it's terrible. Sometimes it's good."[4] Another neighbor of the plant, Carleather Ponder, agreed: "With the Royal Oak plant and the shavings plant, it's awful. . . . The health concerns are just so much. Some days it's hard to breathe when you go outside." [5] Don Williams, another resident of Happiness Homes, said the soot became stuck in the window screens and collected on roof eaves: "This little black stuff, the only place it comes from is burning the charcoal and stuff like that. . . . Sometimes you can smell the smoke in the air," he adds.[6]

**Rude Awakening**

Over the 1980s and the 1990s and into the new century, the protest over soot from the charcoal plant flared up in many households located in Bunche Heights, Richmond Heights, Kings Landing, Happiness Homes, and Busbee Quarters—the neighborhoods surrounding the Royal Oak plant.

Ann Mathis first became aware of the problem one morning when she noticed the flowers in her garden were covered with a white ash that looked like frost. "Everything was white, but when I touched the leaves of the plant, I found they were covered in ash that was black underneath," she recalls. She carried a few of her plants down to Royal Oak and asked managers there what was on her plants. They claimed they had no idea. Then she wrote to the Florida Department of Environmental Protection (DEP) asking them to investigate. Their response was "gibberish," laughs Mathis who, like Reed, is a retired school teacher. "They said the soot was harmless and I should just wash it off with a hose . . . but I knew that was fishy." Royal Oak also manufactured lighter fluid, and the fumes and odors coming from the plant were sometimes so noxious that Mathis once called the police.

"The Royal Oak plant was hazardous to our health and the damage has been done," observes her husband, Plato Mathis, who built their large, handsomely sited home thirty-four years ago across the street from the twenty-seven-hole Pine Oaks Golf Course that covers over a hundred acres. His wife suffers from respiratory distress and underwent heart surgery that they both believe was caused by or at least aggravated by pollution from the plant.

"The town [Ocala] is run by five people—big businessmen and horse people who live on the east side. They did what they wanted to do," reflects Plato Mathis, 77, who worked at a medical center for fifty years before retiring and opening catering business. Not only did the town fathers decide that it was a good idea to open a charcoal factory on the African American side of town next to his house, they also concluded that it made sense to install an asphalt plant and a car smashing and shredding plant that to this day makes unbelievably loud noises. "These city officials always had a place for these dirty factories and it was our backyard," says Mathis.

But local officials went too far when they decided to sell the town-owned golf course to Sysco Systems, a firm that planned to open a food distribution center that would have serviced 350 semitractor-trailer trucks a day pulling into the plant across the street from the Mathis property. In 1996 the Mathises joined with the Reeds and other members of the community to fight and defeat this locally unwanted land use. They also ran a successful campaign to require a recycling plant to remove a mountain of debris piled up behind their house. These early organizing victories paved the way to take on Royal Oak and gave residents a sense that if they stayed organized, they could improve the quality of life in their neighborhood by shutting down some of the polluters and preventing others from moving in.

**Getting Organized**

Not until February 2003 did Ruth Reed, the energizing force behind the antisoot campaign, decided to devote herself full time to shining a spotlight on the neighborhood pollution problem. "We didn't have much money, but we did have a lot of time," notes Reed who was retired. Gathering a couple of dozen concerned citizens in her two-story house, she initiated a discussion about what could be done to help clear the air. At one of these meetings, the Neighborhood Citizens of Northwest Ocala (NCNWO) was organized. In all there were about fifty active members, with some seven core people who did the hard work. The grassroots group's tactics were not confrontational, and there were no protests or marches with residents carrying picket signs. "We wanted to be peaceful," one member explains. Instead they held meetings to increase awareness of the pollution problem, contacted city officials, wrote letters to regulatory agencies, took photos of smoke coming out of the plant's chimneys, captured soot samples, and held fundraising events, including bake sales and car washes. But they were unable to raise enough cash to hire a lawyer.

Help came in the form of Jeanne Zokovitch, an environmental lawyer with WildLaw, a nonprofit that helps grassroots groups with toxic problems. Zokovitch knew personally about what it means to choke on soot: her father, who had worked in the coal mines of Pennsylvania, had died of black lung disease.[7] After visiting Ocala, it didn't take long for Zokovitch to recognize that something was wrong in Bunche Heights and that the density of the smoke coming from the charcoal factory's chimneys was not legal under the Clean Air Act.

Health problems associated with pollution from charcoal factories were well known to the federal regulators, who had faced the issue on a large scale in the Ozark Mountains of Missouri where a large number of charcoal plants are concentrated. In 1997, the U.S. Environmental Protection Agency (EPA) issued a warning about the health dangers associated with the inhalation of very fine particles that result from combustion. These particles, such as the soot from the charcoal plants, lodge deep in the interstices of the lungs, where the body is unable to get rid of them. The EPA "linked the tiny particles to acute respiratory distress, reduced lung function, increased childhood illness and aggravated asthma."[8]

Given the gravity of the health threat, the EPA enforced its regulations on emissions from charcoal plants in Missouri that violated provisions of the Clean Air Act and fined Royal Oak $750,000 in 1997. As part of the settlement, Royal Oak agreed to install afterburners in their Missouri plant that eliminated 80 percent of the fine particle pollution. With this cautionary tale in front of them, other charcoal plants in the state followed suit and installed afterburners to avoid large fines. Once the afterburners were in place, emissions from the plants changed from black to transparent, making a huge improvement in surrounding air quality. But Royal Oak was slow to learn this lesson. Two years later, in 1999, the EPA fined the company another $450,000 for pollution from another of its charcoal plants in Kenbridge, Virginia.[9]

In Ocala, no improvements were made because there was little state regulatory pressure to upgrade pollution controls at the relatively small and isolated Royal Oak furnace despite complaints from residents about the thick black smoke spewing out of the plant's smokestacks. Noticing the discrepancy between what the plant operators were saying about their emissions and what people on the ground next door were experiencing, Zokovitch urged Reed and other members of the NCNWO to recruit more members and raise the profile of the issue in their community.

**Taking on City Hall**

To this end, Reed, the president and indefatigable "stone roller" for NCNWO, repeatedly approached city council members about the soot problems in her community until the council agreed to hold a meeting on the issue in September 2005. Reed and other concerned residents packed the council chambers and presented their petition. City officials were impressed with their presentation and the number of community members who showed up and said they would look into the matter. "We never heard of your group before . . . you people just came out of the woodwork," one of them told Reed.

Although the presentation before the city council was a good first step, Reed understood she needed to build support on the city council to prompt any concrete action. So she and her colleagues decided to tackle city council members one at a time to convince them to do something to reduce the soot emissions. State regulators, who should have been controlling emissions from Royal Oak, were "not doing their job, so we decided we would have to do it for them," she said.

Reed was nervous before her first meeting with a city council member who was a banker. "I explained our problem and he was surprisingly positive," she remembers. He explained that she would have to get her community organized, speak up at the city council meetings, and get more petitions signed. The next council member Reed approached, an insurance broker, was sympathetic to NCNWO's goals but would not commit himself on the issue unless other members were in agreement. And another council member was outright unsympathetic: "What are we supposed to do? Are we not supposed to have any industries in the city?" he asked her. "It's either you [the residents near the plants] or them [the businesses]. We're tired of hearing about these environmental complaints," Reed reports he told her. Other disparaging comments came from officials who thought west Ocala was a reasonable site for these industries because "there isn't much of a community up in that area."

State regulators also showed little enthusiasm for controlling the soot gushing out of the Royal Oak chimneys. "The Florida Department of Environmental Protection didn't care about us because we were a black community on the wrong side of the tracks. We weren't upscale enough to matter," Reed charges, clearly angry at what she sees as regulatory negligence and racial bias. Nor was help available from tort lawyers. Attorneys from out of state who came to look into the possibility of a suit against Royal Oak declined to take the case because they said "we were not financially viable," she recalls. "We were just too poor for them to see it as worth their while to take the case," she adds.

With no help from state regulators or lawyers willing to sue Royal Oak for a percentage of a future settlement, Reed and NCNWO faced an uphill battle. Efforts to meet with the managers of Royal Oak had been rebuffed, so there seemed no opening to convince the owners to voluntarily install better pollution control equipment. Nor was there any help from state regulators who had monitored the plant's emissions in 1996 and determined that they "posed no health threat." Local residents knew that the reason for this was that plant managers had advance warning of the monitoring and shut down plant operations prior to the investigation. "It also matters what you test for and what yardstick you use," Reed observed. "We wanted the air tested again."

### Health Impact

Did the soot from Royal Oak cause an increase in disease in the fenceline neighborhoods near the plant? Many residents are convinced that the fine particulate matter from Royal Oak's smokestacks was responsible for what they describe as a localized epidemic of asthma, respiratory disease, eye problems, and cancer among those who lived on the streets surrounding the facility. Evidence of the extent of community concern over this issue surfaced when 392 residents in the neighborhood who answered a survey listed air contamination as the leading health issue.[10]

Residents in Bunche Heights began to report an increase in respiratory distress in the 1980s, writes Cara Buckley, who chronicled the community's health problems in the *Miami Herald*. Young and old were afflicted. Three-year-old Denzel Deason developed asthma and needed a nebulizer mask to feed oxygen and medicine into his lungs for twenty-minute periods four times a day. Other residents, such as Lily Bell Snow and Jasmine Carter, found their breathing problems let up when they left the neighborhood but intensified whenever they returned home.[11]

Luther Daymon Jr., sixty-four, who lived for twenty-two years in a trailer directly across the street from the Royal Oak, is convinced the plant affected his health and that of his children and sister. "I think they are getting away with causing a hazard to our health," he observes as he sits in his front yard looking at the plant. His sister, Bobby Goodson, who lived across the street from him, died of emphysema at age sixty-six. "She couldn't go out of her door because of the smoke," he recalls. "I think the soot helped kill her," he adds.

By 2006 doctors had hooked Goodson up to an oxygen tank full time. "The tube looped around her ears, piping oxygen up her nose. When the factory was smoking heavily, her children would warn her, 'Momma, don't come outside,'" writes a reporter who visited with her at the time.[12] After his sister died, Daymon consulted with lawyers about a possible suit but none would take his case. "We don't have enough money to fight these plants. They don't put these plants in rich areas; they put them in low-income areas where they zone it so they can come in," he observes.

From his front yard, Daymon, who works as a landscaper and part-time carpenter, has had front row seats from which to view the operation of the Royal Oak facility. He knows the plant inside and out, having once been hired to lay brick and cinderblock at Royal Oak while the owners were refurbishing the facility. He knew some of the men who worked at the plant for $6.50 an hour and would come out after a shift the color of charcoal.

"When the plant was fired up you couldn't be outdoors," says Daymon, who saw flames leaping forty feet in the air above the smokestacks. The soot that came out of the stack "can get into your house even with the windows shut and you know if it can do that, it can get into our bodies also," he notes. He had to give up gardening because pollution from the plant contaminated the soil with chemicals that caused rashes on his hands. "The smoke affected my health, my breathing, my skin, and it burned my eyes" he says, summing up the injuries he suffered.

Regulatory efforts to investigate the operations of the plant were ineffectual because managers always knew when the inspectors would be arriving and would curtail their emissions, he adds. Someone was probably getting paid off, he speculates. But anyone who actually looked around the neighborhood could plainly see it was polluted, he argues. All you had to do was run your hand over the hood of the car, and you would see how much soot had fallen overnight. Often plant operators would fire the furnace at night, and flames from their chimneys would bask his home in a red glow and heat up the side of his trailer. In the morning, the whole neighborhood would be wreathed in gray clouds of smoke that clung close to the ground.

**Busbee Quarters**

Cathy and Calvin Jones witnessed up close the damage done to the health of neighborhood children by the charcoal plant. The Joneses lived in Busbee Quarters, the low-income housing complex of spare wooden cabins located at the foot of Royal Oak's smokestacks.

"No one cares what the kids and all the rest of us in Busbee Quarters were breathing," charges Cathy Jones indignantly. "They restricted us here and then turned it into a dumping ground. The pillars of the white community own stock in the charcoal plants, the low-income housing, and the waste dumps they put in Busbee Quarters where the pollution is killing people. . . . We have a whole generation dying off in these neighborhoods in homes where people have a whole table full of medications. It is murder, torture, and a slow death. This is no secret. These people are committing these crimes openly . . . but the owners of the plants are never prosecuted," she says.

The Busbee Quarters neighborhood is located closer to Royal Oak than any other neighborhood and as a result was more heavily polluted. Residents who lived in the cabins near the plant were reluctant to protest the substandard conditions of air quality because they lived in subsidized units and could be easily evicted from their homes. Similarly, the men who worked at the plant, many of whom had severe respiratory problems, were afraid to complain (or even to consult doctors) for fear of losing their jobs.

"Sometimes the smoke was so bad that you would think that the house next door was on fire," says Calvin Jones, a former member of the U.S. Air Force who had served in Japan. When Jones moved back to Ocala with his wife, he worked first as a corrections officer and subsequently as a tutor at the Evergreen School. Parents of students at the school asked him if he could help the children who lived in Busbee Quarters, where conditions were particularly harsh. Jones talked with school officials about opening a tutoring program in the neighborhood but was told the area was too dangerous (it had a high crime rate and drug dealing was rampant) and that no insurance company would sign a contract on such a venture.

At first, Cathy Jones did not think that moving into Busbee Quarters was a smart move: "We spent a whole lot of energy getting out of places like that," she reminded her husband. But Calvin Jones, who had grown up in a subsidized housing complex, convinced her to give it a try, and they rented a single-story, three-room cottage next to the plant in which they started a home-schooling and after-school program. During the day, they taught children to read, write, do math, and say "thank you, sir" and "thank you, ma'am"; at night they slept on fold-out cots amid the educational materials. "Our idea was that the way out of this poverty was through education," explains Cathy Jones.

The Jones's tutoring center was located in the heart of an active drug trafficking market where young men with guns sometimes forcibly moved into the homes of young single mothers who lived in the subsidized housing. There was little they could do to resist this protracted form of home invasion. "Let me put it this way," says Cathy Jones: "On a good day there was only a stabbing, and on bad days people were killed."

Shortly after he opened the learning center, Calvin Jones had a run-in with two drug dealers who told him to close it down and get off their turf. But Jones, a six foot two inch former middle-linebacker for the University of Florida Gators, was not about to be driven off. By his own account, he returned the next day with a nine millimeter Glock pistol tucked in his pants and a shotgun. "I told them, 'I'm not going anywhere, so either you understand that or we can shoot it out right now.' . . . No one bothered us much after that," he told a reporter.[13] For the next twelve years, the Joneses continued to run their tutoring program in what Cathy Jones describes as "the mouth of the beast."

Giving directions to Busbee Quarters was simple, she recalls: all you had to do was tell visitors to head toward the fire and the smoke. The smoke from the charcoal plant would spew out all night long, and the flames shooting out the chimney would light up the neighborhood. On cool mornings, the smoke would hover close to the ground, creating a gray mist that swirled through the unpaved streets. Trash pickups were infrequent, and mounds of discarded furniture and mattresses soaked in the rain, creating a perfect home for rats and other vermin. Broken glass and drug paraphernalia littered the ground. Attempts to wash and dry clothes and clean homes were foiled by the soot that covered everything.

After neighbors joined together to fight pollution from the Royal Oak plant, NCNWO organized a number of health fairs where health workers were brought in to conduct medical screenings. Almost all the children from Busbee Quarters were diagnosed with breathing problems, and some had to have fluid taken out of their lungs, Cathy Jones recalls. Both adults and kids had asthma, she adds.

While operating the learning center in Busbee Quarters, Calvin Jones worked with city officials to improve trash removal, install street lights, and see that truck traffic was rerouted. Gradually conditions improved marginally, and articles began to appear in local newspapers about this pioneering tutoring program. In 2006 a number of local charities built a new school for the Joneses across the road from Busbee Quarters, just a block from Ruth and Leroy Reed's home. Their work continues as they strive to educate one child at a time and show them a way to succeed. "I had help when I was coming up," says Calvin Jones, who was mentored by nuns and Vista volunteers. "It's only right that we help the next generation."

The damage that the Joneses saw done to the health of children in the neighborhood should have been enough to bring in public health officials to deal with the excess of respiratory disease. But no formal health study has been conducted by any government agencies. However, a community health assessment survey carried out by students from the University of Florida at Gainesville, under the direction of Laura Chen and Susan Glotfelty, did find some disturbing results. For example, among the 80 households that answered the survey passed out to 438 homes, 53 percent said they had witnessed visible ash, 33 percent reported bronchitis, and 22.5 percent had a family member taken to the hospital due to respiratory distress. These results, while preliminary, suggested that a more comprehensive study was warranted. Cathy Jones was diagnosed with a blood cancer and had to have her spleen removed. She is convinced that pollution from the plant contributed to her illness, and she is angry about the social forces that created what she describes as a racially segregated dumping ground.

### Reaching Out for Help

In an effort to kick-start more effective organizing, in 2005, WildLaw attorney Jeanne Zokovitch introduced Reed to Denny Larson and Ruth Breech at Global Community Monitor, who work with grassroots groups confronting contamination problems and provide them with access to both low- and high-tech methods of testing the air in fenceline communities located adjacent to heavily polluting industries. In neighborhoods where the air tests as polluted, Larson and Breech help local groups organize, test air quality, and hold both corporate and regulatory officials accountable.

Reed met Larson when she traveled with Zokovitch to a workshop he gave in Ohio where he explained how groups such as hers could learn to build and operate citizen air monitoring equipment with which they could captured samples of polluted air in plastic buckets equipped with vacuum pumps. These air samples could then be sent in special bags to an EPA-approved laboratory for analysis. This "bucket brigade" technique, used in both domestic and international communities, has proved effective at providing evidence that prompts regulatory action or can be used in lawsuits against companies that pollute.

"Denny just turned on a light bulb for me," says Reed. "We learned all kinds of things from him. For instance, we learned that it was not just the heavy black smoke that could cause harm but also the white smoke. When I left that workshop, I was determined to become a bee in the collar of those regulators and city officials who were not paying adequate attention to our soot problem."

### Airing Dirty Laundry

Reed quickly picked up some of the organizing techniques Zokovitch and Larson had to offer. In one brainstorming session, residents came up with the idea that airing their dirty laundry in public might help officials grasp the intensity of the pollution problem they faced. To capture graphic evidence of the pollution residents were daily inhaling, Reed and her colleagues hung white sheets on the line for a few days to demonstrate how quickly they were darkened by soot from the plant. They left white pie plates and Tupperware outside their homes to collect the soot. Then when their turn came to make a presentation at the city council, they arrived waving the dirty sheets and passing around the pie plates filled with soot so that officials could no longer deny what was plain to see. In the process, the grassroots activists from west Ocala made it painfully obvious to officials that making charcoal was a dirty business and the people who lived closest to the plant were paying a steep price for regulatory inaction.

Residents had also begun to keep pollution log books in which they entered the date, hour, and location at which they saw black smoke spilling from Royal Oak's smokestacks, the odor they smelled, and the impact the emissions had on their health. With these citizen air-monitoring logs, residents began to do the work that state regulators should have been conducting all along: they monitored emissions from the plant and kept records.

Also helpful in raising awareness about the seriousness of the soot problem in Bunche Heights was a sophomore student, Ollie McLean, from Forest High School, located on the predominantly white side of town. She smeared ten white pieces of poster board with Vaseline and mounted five of them within a quarter of a mile from the plant in northwest Ocala and the other five in the more affluent southwest quadrant of the city where she lived, two miles from the charcoal factory. After leaving the poster boards in place for a week, McLean compared them. Not surprisingly, she found that those mounted close to the plant were covered in black soot trapped in the Vaseline, while those placed at a greater distance from the plant were still lily white. As part of her biology science project, McLean calculated that the ratio of soot trapped on the poster boards near the plant was 116 to 1 when compared with those placed farther away.[14]

Faced with the dirty sheets, the soot in the pie plates, and the blackened poster board science project, local officials had to concede there was a problem. "I had no idea of the level of exhaust [coming from the plant].

If this kind of activity was occurring in a more affluent area of town, it wouldn't be 24 hours before something was done about it," councilman Kyle Kay told a reporter.[15] In August 2005 the Ocala City Council passed a resolution supporting Reed's coalition's demands that something be done to monitor the air in her community. Council members began a letter-writing campaign to federal, state, and county officials requesting that the plant be more carefully investigated.

### Missing Afterburners

While she was grateful for the city council's efforts, Reed and members of NCNWO announced that they would go ahead with their own plans to monitor the air. This announcement prompted state officials to act, and in September 2005, the Florida DEP dispatched investigators to inspect the Royal Oak plant.

Previous regulatory visits had repeatedly found the charcoal plant "in compliance" with state and federal air pollution regulations. At their strictest, DEP officials issued warnings to plant managers requiring them to improve the facility's emissions control performance. But Zokovitch was skeptical: for years DEP had relied on reports prepared by a "specially trained" Royal Oak employee who would step outside the plant offices, conduct a visual inspection of exhaust from the smokestacks, and then report to the regulatory agency that everything was fine. "If you look at the record they are in compliance. But if you don't trust the form of monitoring are they truly in compliance?" Zokovitch asked a reporter. Furthermore, Royal Oak had a long history of increasing production late at night, making it unlikely that DEP officials based in Orlando would come all the way to Ocala at 2:00 a.m. to conduct an air emissions test, she adds.[16]

In August 2005, in the face of accumulating questions about Royal Oak's environmental record, Royal Oak spokesman Robert Lockett insisted that air emissions tests done at the plant had demonstrated that the plant was abiding by clean air and environmental laws: "In general I can tell you that we do not only meet the DEP guidelines down there, we strive to do better than them."[17]

A month later, DEP field inspectors found the plant was emitting nine times as much methanol as was legally permitted. Methanol is a neurotoxic chemical that causes dizziness, headaches, irritation of the nose and throat, vomiting, and, in large doses, blindness, convulsions, and death.[18] Royal Oak had reported to the U.S. EPA headquarters in Washington that it was releasing ninety-three to ninety-seven tons a year of methanol into the air, while at the same time insisting to state regulators that it was not a major source of air pollutants. This statement was at odds with the fact that anything over ten tons of methanol releases qualifies a company for status as a major source of air contaminants.[19]

Inspectors also asked to see the charcoal plant's afterburners (designed to reduce particulate pollution) with which corporate officials claimed the plant was equipped. Following a diligent search of Royal Oak's property, inspectors found no afterburners, which are immense pieces of equipment that could not have been overlooked. Inspectors also reported evidence of excessive particulate emissions from the plant's smokestacks. These violations eventually resulted in a fine of $39,500 and a requirement that plant owners begin soil and groundwater sampling as well as dredging a retention pond where contaminants could be contained and water evaporated.[20]

On November 29, 2005, the DEP made public the fact that its inspection had uncovered nine possible violations of the law. The next day, faced with fines or the prospect of being required to spend large sums to install pollution control equipment, officials at Royal Oak Enterprises headquartered in Roswell, Georgia, decided to close the plant and lay off all forty-three employees. In explaining their decision to close the plant, the owners of the nation's second largest charcoal briquettes facilities cited high operating costs, the poor quality of raw materials, and environmental issues at the Ocala plant.[21]

Told of the decision by a reporter, Reed exclaimed: "Hallelujah! David has defeated Goliath."[22] City councilman Kyle Kay echoed this sentiment: "This is fundamentally a good thing for residents around the Royal Oak facility. I am happy for them. It's a new day and they won't have to deal with the environmental concerns of Royal Oak."[23] But Reed was also quick to note that the victory was bittersweet. "We are happy that neighbors will not have to deal with the environmental nuisance any more," she told a reporter. But the closing of the plant came with real costs: "We are sorry for the 43 people who are out of work there. We tried to sit down at the table with Royal Oak to reach some sort of solution. It seems to me they took the easy way out and decided not to work things out with the neighbors."[24]

## Capturing Evidence

After learning that Royal Oak would close, Reed was faced with a new dilemma. Little monitoring had been done of the air in her community that she and her colleagues could use to prove that their neighborhood had been polluted for decades by Royal Oak. This kind of monitoring data might prove essential if the company was ever to be held legally responsible for the alleged damage it had done to the health of community residents, their property, and their quality of life.

Some efforts were made to see if state air monitoring equipment could be brought in to do testing while there were still pollutants in the air, but before that could occur, Royal Oak tore down the smokestacks. Efforts to mount air monitoring devices near the plant were also hampered by the state's insistence that NCNWO or the city take financial liability for the equipment in the event that it was vandalized or stolen. With no agreement in sight about how to mount air monitors, Denny Larson from Global Community Monitor taught Reed how to take "wipe samples," which involves gathering evidence of pollution from various surfaces using a swab to pick up some of the soot deposited by emissions from the plant. An analysis of these samples could detect Royal Oak's chemical fingerprint.

Wipe samples are not ideal for picking up traces of polynuclear aromatic hydrocarbons (PAHs). The preferred method uses air monitoring equipment that sucks in large volumes of air and traps pollutants in a filter. But this method was too expensive for local residents, so they decided to employ the cheaper but nevertheless valid wipe sampling method. Wipe samples were taken of soot that had been allowed to accumulate for six days from the hood of a car parked at the Reed's home and at several other homes close to the plant. They were then mailed to Columbia Analytic Services, a U.S. EPA-certified laboratory for analysis.[25]

Wilma Subra, a chemist who helps grassroots groups with technical issues, was at first skeptical that wipe samples would be able to capture the full range of PAHs emitted by the plant. However, when sampling results came back from the lab, they not only documented Royal Oak's fingerprint, they also found trace evidence of all twenty PAHs for which the laboratory tested. Seven of the chemicals found in the samples are listed as probable human carcinogens by the EPA, Subra notes.

"The health impacts [of PAHs] consist of skin irritation, irritation and burning of the eyes, damages to developing fetuses, birth defects and reproductive impacts," she said. "Even though PAHs are present at below the EPA screening levels, care should be taken to reduce exposure from inhalation and skin contact with the soil and particles," she added.[26] Subra described the presence of so many different PAHs in the soil and homes of west side Ocala residents as "definitely a cause for concern. It indicates the need to do additional testing to determine the extent of airborne contamination."[27]

Subra believes that the charcoal dust in the homes and in the ductwork of the air-conditioning units "has the potential to serve as a source of contamination for many years to come" despite the fact that the Royal Oak plant is closed. The large number of chemicals present in the dust can have a harmful cumulative effect and cause acute health problems, particularly when the dust is disturbed, she notes. "Thus, even though the facility is shut down, the sources of exposure still exist in the yards and homes of the nearby residential area. This exposure, compounded by the exposure experience while the facility operated, could have a long lasting negative impact on the health of the community," she concludes.[28]

County health officials disagreed. Air pollution tests conducted on a vacant block near the plant and inside the homes of several residents determined "no lingering effects from Royal Oak at this point," reports Tom Moore, Marion County's environmental health director. But these county tests were flawed, Reed argues. First, they were carried out after Royal Oak had closed its furnace and torn down its smokestacks; second, they did not address the PAH pollution found in the wipe samples; and third, the county's indoor air-monitoring canisters were ineffectual because inspectors failed to equip them with pumps.

Following the closure of the plant, concerns remain about the possibility that the soil immediately surrounding the charcoal plant on Royal Oak property is heavily contaminated and will constitute a health hazard for years to come. Royal Oak owners were offered a chance to have the soil tested under a federal brownfields restoration program but declined the opportunity.[29] The fate of the property remains unclear.

### Persistent Grassroots Organizing

The victory of environmental justice activists in the struggle surrounding the charcoal plant in Ocala demonstrates that focused and persistent grassroots organizing can work even against large corporations. While the environmental justice victory in Ocala was the result of the work of many individuals, the energy and determination of Ruth Reed was critical to keep the issue simmering during a protracted struggle. "Mrs. Reed is a real go-getter," attests Arthur Spencer White, a member of NCNWO who lives in the Happiness Homes neighborhood not far from the Royal Oak plant. "She built that organization up from nothing and it wasn't easy."

Without Reed, the soot would likely still be raining down on the community, agrees Krista Fordham, forty-eight, another resident who lives within sight of the plant. "It was a one-woman show," observes Fordham, a former television newscaster, who lives with her mother and ten year-old, wheelchair-bound son with a spinal birth defect. "Thank God for Mrs. Reed. People around here tried for years to get some regulatory relief with little result. It wasn't until she became involved that things began to change. Her passion was so evident that it became clear she would not give up until something was done," Fordham observes.

Decades of work as a school teacher provided Reed with skills that served her well as an organizer: she knew how to stand up and speak in front of groups of people and how to get people working together. "She also reached out to people [outside the community] who could help us and brought them in," says Arthur Spencer White, a former Air Force veteran and retired post office employee who lived near the charcoal plant for twenty-five years. "She showed us that you can't just sit idly by while the soot is affecting people in the whole neighborhood. You have to organize. City officials didn't pay any attention to Robert J. Brown when he was the only one protesting. But when we started meeting together on a regular basis at the Church of God and at Howard Academy and we put forty to fifty people together, then they had to start listening. Ruth made that happen," he adds. Reed's organizing work in Ocala stirred up enough consternation about dangerous air pollutants released by the Royal Oak charcoal plant that eventually city officials, journalists, and finally state officials had to take notice, and the plant was eventually shut down.

A winning formula for grassroots fenceline groups such as the one that came together in Ocala is difficult to pin down since circumstances differ from one site to another. For example, if the source of pollution in Ocala had been a giant refinery instead of a small charcoal plant, the chances are it would not have been shut down. Nevertheless, successful campaigns have some common elements. First, a critical mass of local residents has to conclude that a pollution problem exists and needs to be addressed. Then, a small core group of highly motivated residents needs to take a leadership role in organizing the community. These local organizers must be persistent in their efforts—sometimes over a period of years or decades. Media coverage of the problem is often needed to put pressure on state officials and corporate officers to reduce emissions. Help from outside experts is often required to navigate the regulatory process and translate complex reports about chemical releases and exposure rates into lay terms. Residents often must learn to do their own monitoring to capture evidence that they are being exposed to dangerous levels of chemicals. Health surveys can offer anecdotal evidence of a pollution-induced health impact. And organizers must bring residents together to discuss strategy, mount protests, pack city government and regulatory hearings, and lobby officials for regulatory relief. All this takes time, patience, cunning, perseverance, and, at times, courage.

## Civil Rights Issue

At their core, environmental justice struggles such as the one in Ocala are a continuation of the civil rights movement because they confront racial and economic injustices that result in people of color and low-income residents being disproportionately exposed to the toxic by-products of heavy industry. As David Camacho writes in *Environmental Injustices, Political Struggles: Race Class, and the Environment*, "Much of the success of the environmental justice movement in the United States can be attributed to the ability of the movement to align itself to the legacy of the civil rights movement."[30]

The struggle in Ocala fit this mold. Residents, forced to play host to heavily polluting industries, inhabited African American neighborhoods established decades ago when racial segregation was still enforced. "People in my neighborhood live where they [whites] placed them," one resident recalls. In the past, African Americans were not permitted to settle in white neighborhoods and instead were able to buy homes only in west Ocala, explains Ruth Reed, who came to the area from Kansas in the mid-1960s.

The city was racially divided along Highway 441 that separated the white eastern section of the city from the black western area. "At the time, the schools had been desegregated, but the businesses hadn't," Reed recalls. There was an unspoken code about which establishments African Americans could enter. "Restaurants that had 'white' in their name, like the White Kettle, were reserved for whites," she continues. Ocala's African American community finally revolted against not being allowed to eat at local lunch counters, and an economic boycott in the 1960s forced white restaurant and other businesses owners to open their doors to blacks.

What did not change was the pattern of placing highly polluting facilities in the African American neighborhoods. Under Jim Crow rules, minorities were directed to build outside the town limits in areas that were considered undesirable, Reed recalls. The city did not invest much in the infrastructure of these outlying areas, and for years these black enclaves had no street lights, sidewalks, trash pickup, or even water lines. Over time, as the population grew, some of these neighborhoods were incorporated into the city and water lines replaced shallow wells. Residents in these areas hoped they would begin to see their tax dollars spent on fixing up their neighborhood to bring it up to the standard of white districts but the city fathers saw the area as an underdeveloped location that was appropriate as a site for a variety of noxious industries that whites would never allow to be built near them. In other words, they decided to dump all the most toxic industries in the black neighborhoods. As a result, African American families in west Ocala were exposed to high levels of toxic pollution that they claim compromised their health.

The environmental inequities experienced by residents of Ocala are mirrored in an African American neighborhood in Pensacola, Florida, described in the next chapter. In Pensacola, Margaret L. Williams—like Reed, a retired school teacher—reluctantly took on the job of organizing a grassroots antipollution campaign. She fought for years to relocate her own family and hundreds of neighbors who lived in the shadow of a giant mound of excavated wastes laced with the highly toxic contaminant dioxin. It was not by chance, she argues, that two Superfund hazardous waste sites were located in a mixed residential/industrial zone in one of Pensacola's poorest African American neighborhoods. Williams also points out that white residents from Pennsylvania, whose homes were contaminated, were provided with better relocation packages than were the African American families from Pensacola.

# 2

## Pensacola, Florida: Health Problems near "Mount Dioxin" Require Mass Relocation

Margaret L. Williams was raised at 27 East Pearl Street in a house wedged between two heavily polluting factories next to the railroad tracks in Pensacola, Florida. On one side was the Agrico Chemical Company, a chemical fertilizer plant where her father worked. On the other side stood the twenty-six-acre Escambia Treating Company (ETC), where wooden utility poles, railroad ties, and foundation pilings were soaked and pressure-treated with creosote or pentachlorophenol (PCP) to prevent them from rotting. Both facilities have since been declared federal hazardous waste Superfund sites.

Pollution from the plants coated everything, Williams recalls. The screens on the windows of her childhood home were caked in yellow sulfur released from the fertilizer plant, and water from local wells was oily. On some days while walking to school, Williams had to shield her eyes with her hands and cover her nose and mouth with a handkerchief against the dust and odors. Worst of all, her home was like an oven in the summer because her parents closed the windows to keep out the strong chemical fumes.

Flooding was also a problem. The wood treatment plant failed to install drainage pipes that would divert stormwater runoff from carrying pollution from the grounds of their facility downhill into neighborhood yards and houses. Some former employees of ETC tell of being dispatched to Rosewood Terrace, a neighborhood adjacent to the plant, after a heavy downpour to "pump out creosote and PCP that had pooled in the yards and distribute sand over the contaminated areas," Williams reports.[1]

In the 1940s and 1950s her father had it worst of all. He worked in the bag room at Agrico where chemical fertilizers were packaged for shipment. Management did not provide its employees with respirators, goggles, or boots to protect them from the dust and chemicals, so her father took his own precautions. He wore a bandanna over his nose and mouth and wrapped burlap bags over his shoes to keep the chemicals on the floor from damaging his feet.

Burlap bags were also used as protective footwear at the Escambia Treating Company plant. One former worker described climbing in the long troughs filled with creosote and spinning the poles with his feet until the wood was completely saturated in the black tarlike substance that keeps them from rotting. After a batch was finished, the wastewater contaminated with creosote was dumped into an unlined hole in the ground where it seeped into the groundwater over the years. Residents in the surrounding homes, who depended on shallow wells, had to pump the water for several minutes to clear the oil substance before getting relatively clean water. Years later, their wells were capped by county officials and city water was piped in. "No one told us at the time that the reason they capped the wells was that the groundwater was contaminated," Williams notes. "We just thought they were improving the system."

**Pouring Wastes in an Earthen Pit**

A U.S. EPA document describes the process at ETC in some detail. Starting in 1944, yellow pine logs were shipped to the plant, where they were debarked, formed, dried, pressure-impregnated with preservatives, and stored before being shipped. From 1944 to 1970 coal tar creosote (the black, sticky stuff on utility poles) was used. In 1963, ETC started experimenting with a new wood preservative, PCP, a strong biocide that kills fungi, insects, and marine organisms that can destroy untreated wood. PCP is often contaminated with dioxin, one of the most highly toxic chemicals in the world. From 1970 until 1982, when the plant closed, PCP was used in place of creosote.

From the 1940s to the mid-1950s contaminated wastewater from the plant was sent to an unlined earthen impoundment area—a hole in the ground. After the mid-1950s an oil/water separator was installed to recapture some of the preservatives. A hot pond of contaminated wastewater was sprayed by giant showerheads into a cold pond. In the process, some of the organic constituents of the waste turned into vapor.

The water that was separated out was dumped into the Pensacola sewer system, while the contaminated liquid runoff was diverted into a series of concrete separation basins where some of the wood-treating preservatives were recaptured for reuse. Contaminated wastewater from this stream was shunted to an impoundment known as "the swimming pool," which had a capacity of 225,000 gallons. There the hot Florida sunshine evaporated some of the contents before the rest was flushed into the sewer system. This waste management system continued in this fashion for over forty years before the EPA finally initiated inspections and began to cite ETC for violations in 1985.[2]

"Wood preservative treatment facilities have contributed greatly to the ranks of Superfund sites," note experts at Beyond Pesticides, an environmental group based in Washington, D.C., that has long warned of the danger to human health and the environment of pesticides and promoted alternatives to their use. PCP has been found in 314 of the 1,300 Superfund sites. "So much illness has resulted from worker exposure to pentachlorophenol that it is seen as a significant source of income for attorneys pursuing toxic torts," writes the group's executive director, Jay Feldman.[3]

## Health Impact

The ground surrounding the homes of Margaret Williams and her family was later found to be heavily polluted with dioxin and polynuclear aromatic hydrocarbons (PAHs), the banned pesticide dieldrin, and arsenic. In fact, the contamination in her neighborhood was the worst ever found off-site next to a wood treatment plant, notes Wilma Subra, a chemist who works with grassroots groups dealing with contamination problems. "This is the worst wood treating site for impacts to public health," she says. Blood sampling of former ETC workers and residents who lived near the plant were found to have "elevated levels of dioxin in their blood in excess of the general population" twenty-five years after the plant closed, Subra observes.[4]

While it is impossible to prove that the large number of illnesses and early deaths in Williams's family were caused by exposure to these chemicals, it does seem likely. Both Williams's mother and father died of cancer, as did two of her uncles. The family suffered from respiratory problems as well: one of her uncles had to be taken out on the porch every evening and thumped on the back to clear his lungs. Williams's daughter recalls that her aunt had an enlarged tumor that made it difficult for her to sit and her life a misery. It was hard to find a house in the neighborhood where someone in the family had not died of cancer, she adds.

The size of the cancer cluster in the neighborhoods near the plants was never tallied by professionals, but anecdotal evidence suggests it was substantial. Lisa Wiggins, twenty-seven, who also lived on East Pearl Street with her three children, knew many residents in her immediate neighborhood who died of cancer. "The lady across the street, her mama died of cancer. The man up the street, him and his brother died of cancer. The lady on the corner, she died of cancer. . . . I know if we stay here, exposed to these kinds of chemicals, I mean we are not going to have a chance," she told a reporter.[5]

Reproductive disorders were also widely experienced by women in the Rosewood Terrace, Oak Park, Goulding, and Clarinda Triangle neighborhoods near the plants. Williams's mother had four miscarriages before she finally gave birth to a daughter who lived. Williams, in turn, had two children who died—one a stillbirth and the other who died of respiratory problems three months after birth. "Many of the young girls had hysterectomies before they were twenty, notes Williams. Her own daughter was one of them.

### Opening Pandora's Box

The capping of contaminated wells in Williams's neighborhood was an early sign of deeper problems. The ETC site is located on a high point in Pensacola atop a shallow sand and gravel aquifer, which made it a perfect source point for a large plume of contaminated groundwater that to date remains unremediated. The U.S. EPA now recognizes that the wood treatment plant contaminated ninety-six acres of "impacted land" and generated a plume of contaminated groundwater that extends 1.3 miles downhill to Bayou Texar.[6]

The size of the contaminated area is hardly surprising given that "few environmental precautions were taken," during the wood treatment plant's nearly forty years of operation, observes Williams. The owners of ETC, including the former mayor of Pensacola, likely knew that regulators were closing in and that a day of reckoning was at hand. It appears more than coincidence that just before the plant closed in 1982, the owners sold it to their workers. The company then declared bankruptcy, leaving it to taxpayers to foot the bill for the cleanup.

ETC was abandoned in a deplorable condition with "leaking and unlabeled drums, a lab full of broken equipment and opened containers, an overturned electrical transformer, crumbling asbestos insulation around a boiler—as well as soil, sludge, and groundwater contamination from the waste pits," Williams reports. The levels of contaminants found on site were staggering. The soil was contaminated with as much as 1.09 parts per million (ppm) of dioxin, 545,000 times the acceptable residential limit. Dioxin is known to cause or promote cancer in humans. In addition, there were high levels of other highly toxic contaminants including creosote, pentachlorophenol, furans, naphthalene, PCBs asbestos, benzene, toluene, xylene, chromium, and dieldrin. For years following the plant's closing, adults in search of salvageable construction materials and children looking for fun wandered through the abandoned and contaminated facility that was only partially secured by a broken chainlink fence.[7]

In September 1991, nine years after the company had been abandoned, EPA inspectors found elevated levels of PCP in the subsoil of the site. Soil samples revealed 160 to 180 ppm of PCP at four-foot, eight-foot, and fourteen-foot depths. In October 1991, when the EPA finally began digging out the contamination at ETC, they pried the lid off Pandora's box. By April some 54,000 cubic yards of waste sludge and soil had been excavated and piled up. A memo from the contractor hired to do the remediation work said the sludge could be as deep as forty feet in the ground and estimated the total volume to be over 100,000 cubic yards. But that was not the half of it: four months later, that estimate was increased to 180,000 cubic yards and by January 6, 1993, it rose again to 255,000 cubic yards.

By then contractors had encountered monumental difficulties. Severe weather had hindered efforts to install a stormwater drainage and erosion control system. In other words, storms were washing contaminated soils into the surrounding residential areas. A storm in 1992 with winds up to 60 mph had destroyed the temporary cover on the stockpiled wastes and damaged part of what was to become the temporary tarp that covers the waste today. How much dioxin-laced dust this storm and the others that followed scattered over the neighborhood is unknown. By the spring of 1993, the permanent plastic cover was in place.

There is no question that EPA officials seriously miscalculated the size of contamination problem at ETC and failed to adequately protect residents living next to the excavated sludge, Williams argues. Furthermore, the agency decision to do the work under "emergency removal" authority instead of proceeding with a normal remediation program was, in Williams's view, "a serious mistake." Under the emergency removal provision of the Superfund law, government contractors are permitted to start work without providing public notice, and there are no provisions for residents to participate in decisions that may affect their health. "It was like martial law. This was a quick and cheap cleanup effort. They thought they could get by with this in an African American community in Pensacola," charges Frances Dunham, a local antipollution activist who became one of the EPA's most vocal and persistent critics.

### Rude Awakening

The predicament that Williams and her neighbors faced when they learned about their exposure to chemicals from the two neighboring Superfund hazardous waste sites located next door is not uncommon in communities of color. A study in 1987 found that three of five African American and Latino residents have illegal or abandoned hazardous waste dumps in their neighborhood, and a follow-up study in 1994 revealed that people of color were "47 percent more likely than Whites to live near these potentially health-threatening facilities."[8]

Despite this established pattern, over and over again one hears from residents about their surprise and outrage at waking up one morning to find "men in moonsuits" excavating large volumes of contaminated soil within sight of their yards. The ETC plant had been abandoned for years, and residents had no warning that the dig was about to start. State officials too were caught off guard by the EPA's unannounced excavation. Even Ed Middleswart, a former district air program administrator for the Florida Department of Environmental Protection (DEP), was unaware that the EPA cleanup of the site had begun. He first heard about it "when we began to get calls, blitzes of calls, about people moving around in moon suits, earth moving equipment and creating a lot of dust and odor problems in this little neighborhood. . . . We found out it was the EPA and they weren't telling us what they were doing or why. We were just in the way," he told a reporter while clearly in a state of pique.[9]

EPA officials had known that the groundwater was contaminated since 1987 and about violations at the plant since 1985, Dunham points out. They should have temporarily relocated residents to safety before they began to dig (or burn) the contaminated soil, she asserts: "Instead you had workers in full protective gear excavating contaminated soil while only fifteen yards away children were playing with no protection." EPA officials would later determine that some residential areas surrounding the plants—such as Rosewood Terrace, Oak Park, Escambia Arms, and the Pearl Street/Hermann Avenue neighborhoods—had either "unacceptable cancer risks" or "unacceptable non-cancer health risks."[10]

As the digging continued throughout 1992, the area that needed to be excavated kept expanding. To remove the giant volume of contaminated soil, EPA contractors clawed a forty-eight-foot-deep hole into the ground that was located so close to some residential buildings that their foundations began to crumble and had to be shored up. The excavated soils were piled up into a mound sixty feet tall, a thousand feet long, and thirty to forty feet wide. Some estimates suggest that the pile created by EPA contractors now consists of 344,250 tons of contaminated soils.[11] However much it weighed, the giant mound became the tallest feature in Pensacola, and residents began to refer to it as "Mount Dioxin."

**Incomplete Cleanup**

Despite these herculean efforts of EPA contractors, the removal of contaminated soils was never completed. "They didn't stop digging because they ran out of contaminated soil; they stopped because they ran out of money," Dunham explains. EPA officials say the excavation of the site cost some $5 million. Treating the contaminated soils would have cost at least $43 million, more than twice the annual Regional Removal Program budget.[12] Cleanup of the groundwater is expected to cost anywhere from $19 million to $56 million. By the EPA's own estimate, in addition to Mount Dioxin, there remain 50,000 cubic yards of contaminated soil near the pit and beneath it. The new estimate of the amount of soil that needs to be treated adds up to 400,000 cubic yards after contaminated soils from surrounding residential properties are included.

Unable to treat such a large volume of contaminated soil (there was a brief effort to incinerate some of it on site), EPA officials decided to cover it with a sixty millimeter plastic liner. Residents were first told the plastic cover would last for five years, but the EPA subsequently claimed it had a ten-year life span. In 1996, the contractor who installed the cover reported to the EPA that it was damaged and had a two-foot hole and a two-foot tear in it, along with other smaller holes.[13] The covering has now been in place fifteen years, long past its expected lifetime, Dunham points out. Some residents reported that small trees had grown up through the plastic covering. In 1998 repairs were made of the cover, but by 2002, a U.S. Army Corps of Engineers study concluded that the plastic cover was wearing out.[14] Officials were concerned that a catastrophic failure of the covering might cause dioxin-contaminated soils to be blown over a wide area. A further source of resident exposure occurred when local children, who snuck through the fence, discovered that the plastic covering over the contaminated soil served as an excellent slide.

Dunham judges the way in which the EPA handled the excavation a mistake. Not only did EPA officials not solve the problem posed by the contaminated soils, the digging actually accelerated the rate at which the plume of groundwater contamination spread. Failing to find temporary housing for residents during the two years that the digging took place was outrageous, she continues. Even EPA officials concede that the cleanup around ETC has not been their finest hour. "It is clear that it was a major mistake to dig up the material and leave it there in place for this long a period of time. That is not something that we would like to have done," admits EPA deputy assistant administrator Tim Fields.[15]

## Getting Organized

In 1992, while the excavation was in process, residents in Rosewood Terrace, Oak Park, and Goulding (the communities adjacent to the plant) and in Clarinda Triangle (the community across the highway) began to experience a sharp increase in acute respiratory distress, nosebleeds, headaches, nausea, skin rashes, and a host of other ailments. The air had become so laden with dust from the EPA's bulldozing that residents decided they had to do something. Contractors doing the excavating were supposed to keep the dust down by spraying it with water during the excavation, but as one commentator on engineering ethics pointed out, the expense of spraying the water was bound to cut into the contractor's profits.[16]

Concerned by the decline in air quality and the increase in respiratory distress, residents met first at a neighborhood school and subsequently at the New Hope Missionary Baptist Church in March 1992. They decided that organizing themselves into a group would make it easier to lobby local officials to help them stop the excavation. Thus was born Citizens Against Toxic Exposure (CATE).

Margaret Williams was not looking to become an activist: "They offered the presidency to a bunch of other people, but I was the only one foolish enough to accept," says Williams, whom other residents knew as a former teacher who helped local high school graduates find jobs. "I thought once we were organized, it would be no problem, and I could call on some elected officials to get us some relief," she recalls. "I was surprised and amazed that we got no help at that time." Officials who did write letters to the EPA were told the agency was doing the best it could to dig out and neutralize the source of the contamination that was leaking into the groundwater. From the perspective of residents, however, the cleanup itself was exacerbating already deplorable environmental conditions. The remedial excavation was creating clouds of contaminated dust in a heavily populated urban area. Sometimes Williams describes the situation using diplomatic language: "They were affecting people in a negative way while they did the cleanup." At other times she is blunter: "We thought they [EPA officials] were coming in to help us. This recklessness would not have occurred in nonminority or wealthy neighborhoods."[17]

Williams was able to make this statement based on expert technical advice given her by Joel Hirschhorn, a former government employee who worked on Superfund issues for years. Hirschhorn examined voluminous EPA documents and found evidence "that the original removal action had left very high levels of site contamination all over the site including in open pits and the areas not covered by the pile of excavated materials." The remedial work neither removed the threat to shallow groundwater, "given originally by the EPA as the main basis for the action," nor protected residents, he writes. This analysis provided Williams with a basis to contend that the removal action "had itself caused preventable health threats."[18]

Unable to stop the digging, Williams and other CATE activists organized protests, including one at which residents held a mock funeral procession, carrying a casket past the gates of the plant. They also planted forty white crosses in front of homes where a family member had died in the previous five years of cancer or respiratory disease. As residents became more actively involved in the protest, they began to demand relocation: either the EPA should stop the digging, or they should move the people affected, they reasoned.

Dust from the excavation caused a number of problems. One woman said her daughters could not play outside because "the air would make them itch and burn, and give them headaches"; another woman got so dizzy working in her garden that she fell against the wall. Some residents tried to stop the excavation by standing in front of the bulldozers. After continued protests and organizing by CATE, the EPA finally placed the ETC on the National Priorities List (NPL) of thirteen hundred heavily polluted hazardous waste Superfund sites in December 1994.[19] This decision came late, Williams notes, citing a study revealing that it took 20 percent longer for abandoned hazardous waste sites in minority communities to be placed on the NPL list than it did in areas with a higher percentage of whites.[20]

**Reaching Out for Help**

To get her neighbors relocated out of harm's way, Williams realized she would need allies from outside the community who could bring public attention to the problem and embarrass officials into action. To this end, Williams cast a wide net in looking for support. She called the Southern Organizing Committee in Atlanta and contacted Lois M. Gibbs at the Center for Health, Environment, and Justice, who decades earlier found herself in a similar situation at Love Canal, New York, a community also contaminated with dioxin. Gibbs had recently written a book that described in detail the toxicity of dioxin as well as what residents could do to protect themselves.[21] She described the contamination around the ETC wood treatment and Agrico chemical fertilizer plants as "absolutely the worst" site because of the toxicity of the chemicals found there.[22]

Williams also traveled to Washington, D.C., to take her case to Carol Browner, then the administrator of the EPA, who knew about Mount Dioxin from her previous work as a state regulatory official in Florida. When Williams was finally able to meet with assistant administrator of the EPA Elliot Laws, she told him that he needed to reassess the cleanup at Mount Dioxin. Laws assigned Robert Martin, the agency's Superfund ombudsman, to review the remedial work being done at ETC. When Martin toured Rosewood Terrace, adjacent to the cleanup site, he was so dismayed with the extent of the contamination and touched by the widespread human suffering he witnessed that he pushed for testing of the soils in the surrounding neighborhoods and ultimately for a relocation action.

EPA tests found elevated levels of dioxin, dieldrin, benzo(a)pyrenes, and arsenic. The highest level of dioxin found in Rosewood Terrace was 2,956 parts per thousand (ppt), over 400 times the state residential standard of 7 ppt; and the highest level found in Goulding was 125 ppt. These results were alarming enough to warrant relocation. However, the EPA and the U.S. Army Corps of Engineers charged with carrying out the relocation plan took a minimalist approach. In April 1996, the EPA offered to move 66 Rosewood Terrace families. By August they upped that number to include another 35 households in nearby Oak Park, for a total of 101 households.

**The Push for a Better Relocation Offer**

CATE members rejected this opening offer and held out for more families to be moved. Specifically they pushed for including all the inhabitants of Escambia Arms, a low-income public housing project on the fenceline with ETC, as well as residents in Goulding. To bring pressure on the EPA to expand the relocation offer, CATE called a meeting to discuss the issue at the New Hope Missionary Baptist Church where some 350 angry residents showed up on a steamy August evening to tell federal officials what was on their mind. Officials were presented with three hundred petitions calling for a comprehensive relocation plan. "One by one they filed up to the microphones on either side of the sanctuary to voice their objections." Rejecting the agency's claim that additional study would be needed to justify relocation of other residents, the crowd held up hand-lettered placards and chanted: "No more studies—move us now!" A number of residents who were eligible to be moved said they would hold out until their neighbors were included. "If we were white it would be a completely different story," Williams told the crowd.[23]

Others in high office were sympathetic to this perspective. President Bill Clinton had signed an executive order directing federal agencies to do what they could to improve conditions in environmental justice sites where minority residents were disproportionately burdened by pollution. Williams, who had been appointed to the National Environmental Justice Advisory Council (NEJAC), convinced the group to hold a workshop on relocation in Pensacola. Also turning up the pressure on the administration to relocate residents near ETC was Lois Gibbs, who took out a full-page ad in the Florida edition of USA Today on October 1, 1996. The ad featured a photo of children playing near Mount Dioxin. The caption next to the photos quoted President Clinton saying, "No child should ever have to live near a hazardous waste site."

Two days later, on October 3, 1996, the EPA acceded to CATE's demand to move all 358 households living in the shadow of Mount Dioxin. This was a historic breakthrough. The Pensacola relocation initiative, which took place between 1997 and 2005, was the first major relocation by the EPA of African American residents and cost $25.5 million.[24] It was the third largest relocation in EPA history following the one at Love Canal in 1980 and the Times Beach, Missouri, relocation where 2,000 residents were moved in 1982. Both of these relocations involved largely white populations.

## Low-Ball Assessments

Once the decision was made to go ahead with the relocation, Williams worked to ensure that residents who had been harmed by the contamination were "made whole," as relocation law requires. In theory they were supposed to be placed in a home that was as good as or better than the dwelling they previously inhabited. The first round of appraisals, however, yielded low assessments of their homes, ranging from $20,000 to $27,000. When residents demanded that new appraisers be assigned, some of the assessments rose to $40,000 to $70,000. Residents were also eligible for up to $22,500 in additional relocation funds to place them in homes comparable to those they abandoned. Outraged by these low estimates, some residents contacted a local real estate agency to do a comparative market assessment on their home that the U.S. Corps of Engineers had told them was worth $70,000. What they learned was that comparable homes that had recently sold were valued at $134,900 and $135,000.[25] Not only were assessments low for obscure bureaucratic reasons, residents were not permitted to see the full appraisal of their home to verify its accuracy. Instead, they were presented three homes in their price range to choose among. If they found none of them satisfactory, they could try to find one on the open market on their own.

Fear drove some residents to accept an unfairly low payout for their home, Williams contends. Many were concerned about the damage the contamination was doing to their family's health and wanted to leave quickly. As a result, some residents accepted replacement homes that were shabby and located in rundown, high-crime neighborhoods, she asserts. "The plumbing, wiring, and the roofs in those places were no good and some had rotting floorboards," Williams reports. She pointed out a number of other housing defects to officials at the U.S. Army Corps of Engineers, which the EPA had hired to manage the relocation program, but the Army Corps failed to adequately protect home buyers, Williams charges.[26]

Furthermore, the EPA promised residents that no one would have to take on additional debt to move into a home that was equal to or better than the one they were leaving. This promise was not met, Williams argues. A lot of folks ended up spending their savings to fix up homes, bring the utilities up to code, and fix leaking roofs, she observes. To pay for the house repairs, some relocated residents were forced to take out a mortgage, whereas previously they had lived in the home they owned debt free, Williams explains.

Residents were given a bad deal on their replacement homes for a number of reasons, none of them their fault, observes Joel Hirschhorn, CATE's technical advisor. Assessments were low because they were located near two Superfund sites in a heavily industrial part of the city. Furthermore, for historic reasons, the neighborhood was segregated and had previously been one of the only places in Pensacola where African Americans could buy homes. Finally, by the time the appraisals took place, many homes had already been abandoned by families that had moved out because of health problems caused by the pollution. Other homes in the neighborhood were in disrepair because their owners were too sick to maintain them, Hirschhorn points out. All of these factors lowered property values.[27]

Although residents were paid very little for their homes, others were in line to make a lot of money on the land that was opened up as a result of the mass relocation. Ironically the land residents abandoned was much more valuable than the amount they were paid for it. "Someone is going to make a huge profit when this land is sold," one real estate expert opined. A 1995 Relocation Feasibility Study notes that nearby commercial/industrial land was selling for 250 to 300 times the price of residential property.[28] Unfortunately, the people who owned the homes were not going to be the ones to profit from the change in zoning and land use.

### Better Deal for Whites

Federal relocation programs in contaminated fenceline communities are rare and to date have not treated residents of all neighborhoods equally. Residents in more affluent communities are sometimes given better financial compensation packages for relocation than are those in lower-income communities. Not surprisingly, this inequity in treatment gives rise to resentment and charges of bias.

Government officials do not set out to treat rich residents better than poor ones when shaping relocation programs. They simply respond to pressure and market conditions as they try to place people in comparable substitute homes. The problem arises because federal officials have yet to formulate a uniform plan about how to deal with contaminated communities. As a result, decisions are often made at the local or regional level and thus vary considerably from one area to another.

In Pensacola, bitterness intensified over the low assessments when African American residents living near the shuttered wood treatment plant learned that whites in Pennsylvania, who also encountered contamination problems, had been given a better deal. Word spread that preferential treatment had been afforded a group of forty white homeowners who were relocated out of contaminated houses at the Austin Avenue

Radiation Site in Delaware County near the town of Landsdowne outside Philadelphia in suburban Pennsylvania. According to EPA documents, forty households in Delaware County lived in homes built with some materials contaminated with radium, thorium, radon, and asbestos. Decades earlier, radioactive materials had been mixed in with the construction materials used to build these homes. The radioactive contaminants had come from the W. L. Cummings Radium Processing Company, which operated from 1915 to 1925.

The EPA spent large amounts of money to make these homeowners whole. Some eighteen of the forty homes were decontaminated at a cost of $24 million while the residents were placed in temporary housing. Owners of the remaining twenty-two homes were given an option to relocate or have new homes built under a program that cost an additional $31 million. Of these twenty-two home owners, four were not given a choice about the way in which their homes were remediated, eight chose to permanently relocate into comparable homes they found on the market, and ten opted for the EPA to build them replacement homes.[29] The appraised value of these ten homes was an average of $147,000. Instead of offering to buy them existing homes in this price range, as was the case in Pensacola, the EPA agreed to build them new homes that cost an average of $651,700 each. These were custom-built homes designed to replicate the ones they were replacing. One of them, appraised at $200,000, cost the government a staggering $911,411 to rebuild.[30] "That proves that there is pretty strong discrimination involved in these relocation schemes," observes CATE president Margaret Williams.

After reviewing the costs entailed in the Delaware County relocation program, the EPA inspector general admitted that the program had gone too far. "We do not believe the EPA should be in the house building business. Furthermore, the EPA was not mandated to replicate every facet of an existing structure," the report concluded.[31] The inspector general's report also confirms that "due to public and political pressure, the Region allowed most of the owners the option to either relocate or to have new houses built on site."[32]

"Is this just more of EPA's environmental racism?" asks Williams. "We never asked for anything but homes that are 'decent, safe, and sanitary,' as the EPA promised."

## Not a Precedent

While CATE won its struggle with the EPA to have residents relocated, the agency was careful to couch the reasons for the relocation in language designed to avoid setting a precedent. Officials conceded that twenty-one households were moved because of a health threat from exposure to dioxin, but most of the other residents were moved because of a combination of factors that made the site "unique," EPA officials argued. "The uniqueness of the site and the interaction of many factors present here do not, in EPA's opinion, create a precedent for relocation at other superfund sites," agency officials stipulated in their Record of Decision in February 1997. This sounded to some observers like an effort to discourage residents in other contaminated communities from demanding relocations. "The EPA was on track to portray the Escambia site as a special situation to avoid setting any precedents for relocations nationwide," notes CATE's technical advisor Joel Hirschhorn.[33]

## Success for One Holdout

Not all of the residents who lived near the wood treatment plant in Pensacola got a raw deal. Some of Williams's neighbors who refused to be moved into an inferior house ended up with good homes. One of these was sixty-nine-year-old Jean Roshell, who for thirty-two years had lived fifty feet from the fenceline with Escambia Treating Company near the CSX railroad tracks in a blue house. Like Williams, Roshell remembers the bad old days when yellow sulfur caked her windows. Signs that the area was heavily polluted were not hard to detect. Roshell describes a day when county officials came to pump out a nearby well "and it gushed black stuff that burned one of the men," she says. Flooding also spread creosote into her yard until she prevailed on officials to dig a diversion holding pond near their home for the oily runoff. With this history of pollution problems, Roshell refused to accept what she considered an inferior house that authorities offered in a relocation package. "I decided to fight them all the way. I stuck it out so I got the best deal," says Roshell, who now lives in a handsome brick home. But it took a lot of guts to stick up for her rights and negotiate with the authorities.

After all her neighbors were relocated, it was scary living for two years with her grandchildren down by the railroad tracks with no one else around. "When I would come home, I would rush to get in the house and fasten it up," recalls Roshell, who worried about crime and missed her neighbors. It had been a tight community, and people helped each other every day, she explains: "If anyone needed something, I would take them to the store." Authorities turned to intimidation tactics to convince her to relocate, Roshell claims. A police cruiser frequently parked outside her house, and authorities threatened to cut off her water and light, she says. But Roshell refused to budge: "They didn't scare me. My mama taught me not to let anyone push me around."

The leadership of Margaret Williams in the relocation campaign was a great comfort during these difficult days: "Some [residents] were afraid to protest but we weren't. Miss Williams led us. When she spoke you could hear a pin drop. She was intelligent, kind, and spoke well. And she was a great leader who did a lot for our community."

Roshell is convinced that pollution from the ETC plant took a heavy toll on her family. Her husband, retired from the Navy, developed breathing problems and died in 1992 of lung cancer; her mother had died a year earlier from liver cancer, despite the fact that she never smoked or drank alcohol; her son developed skin rashes and was frequently sick; and her daughter's health problems cleared up when she moved away from home, Williams recalls. Furthermore, two of her daughters had premature babies—one weighing one pound three ounces and the other three pounds three ounces. One of her daughters and a grandson were born with developmental delays. Roshell herself had bronchitis and developed skin cancer. The doctor said that it was unusual for an African American woman to get this type of cancer and suggested that it might be a result of living near the chemical plant. Roshell also took in two children of a woman dying of breast cancer she met in the church where she played the organ.

Then, in 1991 "men who were all covered up in white suits" started digging up the soil just across the fenceline from her home. "They didn't tell us anything about what they were doing," she says. "With the digging you couldn't hardly breathe and I hated coming home," Roshell recalls. At that point she started attending CATE meetings.

### Environmental Racism Charged

Residents of Rosewood Terrace, Oak Park, Goulding, and Clarinda Lane did not fail to notice that the most polluted sites in Pensacola near factories and waste treatment plants were in low- to moderate-income African American neighborhoods where some 925 people lived within a quarter mile of the wood treatment plant. In addition, there were five day care centers, one hospital, and three public schools within a mile of the plant.[34] "I think it is racist," observes seventy-year-old Jimmie L. McWaine, who lived a hundred yards from ETC and had played on the grounds surrounding the facility during his childhood. "Seems like they don't care. They know the neighborhood is right there but they just dump the stuff in the ground. If you notice, they don't do that in the white part of the city or county," he points out.

While ETC was still in operation, the creosote process it used caused health problems for years, McWaine explains. The fumes were worst in the summer when the creosote got hot: "It just burned your eyes," he remembers. And flooding brought contaminated muddy water into homes, he adds. Mostly the pollution was invisible, but sometimes the damage that the chemicals were doing was clear. McWaine grew vegetables in a large garden. "Then, all of a sudden nothing would grow: the ground was contaminated," he recalls. McWaine thinks his family's health was damaged by the pollution. His brother-in-law, who worked at the plant, died in his forties of heart failure, his wife died of respiratory and heart disease, and he had "problems breathing" and skin rashes. Regulatory officials were doing nothing to protect the community, he continues: "You've got to be just about dead for the government to take action."

After working for thirty-nine years making nylon at Monsanto and doing other jobs, McWaine was ready to retire "when guys in moonsuits [hazmat protective gear] showed up and we learned our community was contaminated," he says. McWaine was one of many local residents who had spent decades paying off their mortgages and taxes so that they would have a house free of debt when they retired. "I didn't want to relocate. I was comfortable where I was," says McWaine, who loved his neighborhood despite all its problems. Families knew each other, and there were many close friendships; people looked after each other: "If one hurt, the whole community hurt," he explains. But when he learned the extent of the contamination he decided to move. "People were just scattered. I'll just have to start over," he says.

## Clarinda Triangle

On the other side of the highway from Mount Dioxin and the former ETC lies the Clarinda Triangle, a community of fifty-five family homes and 350 people who live in modest brick or wooden houses and trailers located on narrow winding lanes in a mixed residential/commercial area just off heavily traveled Palafox Street. Initially the EPA refused to test soils in the yards of Clarinda Triangle homes, arguing that contaminations would not cross the broad thoroughfare. However, persistent lobbying by CATE and pressure from the Florida DEP eventually convinced federal officials to test the soil. They found dioxin levels as high as or higher than those found in a number of the neighborhoods closer to ETC. As a result of these findings, a second wave of relocations is now underway in the Clarinda Triangle neighborhood.

"We drank the well water and bathed in it, but it would leave a funny odor on your skin," recalls fifty-year-old Deborah Anderson, who lives in a solid brick house her grandfather built in Clarinda Triangle in the 1960s. Anderson has lived in this house all her life, married a man who worked at ETC, and raised sixteen children. Two of her children were premature, and one was born with spina bifida. Her children also suffered from constant rashes, irritated eyes, and a variety of other health problems.

When testing by the EPA showed that the dirt in her yard was heavily contaminated with dioxin and other chemicals, Anderson knew who to blame. "It's Escambia Treating Company that is responsible," she charges. Recently widowed, Anderson's husband, who worked at the plant, died of lung cancer at the age of fifty-three. For years he had stood on the utility poles that were soaking in creosote with nothing on his feet but sneakers. Eventually he developed crystals in his feet and died from inhaling the fumes, she asserts. There was no compensation from the company after he died. "The government wasn't there to help us either," she adds.

Anderson used to visit friends who lived in Escambia Arms, a subsidized housing project across the highway and just opposite the wood treatment plant. The children in the complex played in the dirt and rain-water near the plant and then started "sledding" down the tarp over Mount Dioxin, she recalls. Adults tried to keep them out, but they would break through the fence. "Many of them had horrible rashes and lesions on their skin," Anderson recalls and then observes, "I don't think they would allow that to happen in a white community."

Clarinda Triangle residents tried to get in on the relocation deal offered to residents across the highway but were left out. It was at that point that Fred Weatherspoon organized the Clarinda Triangle Association to lobby for relocation, and Anderson joined up. "A lot of people were dying, and whole families wanted to move out," she recalls. Her oldest son who had already moved tried to convince her to leave: "Mama, you don't need to be staying in that contaminated area," he told her. But Anderson needs financial help to move.

### "We Will Be Relocated"

"It's terrible to learn that you have been living in a cesspool of contamination," says Katherine D. Wade, the forty-three-year-old president of the Clarinda Triangle Association. Wade, who lived with her family for the past eighteen years in a wooden house not far from the highway, says her work with the developmentally disabled and child support enforcement "taught me to be strong and how to advocate for the underdog." These attributes turned out to be useful as neighbors turned to her to lead the relocation effort.

Wade is both an activist and a budding diplomat. Many residents of Clarinda Triangle, who were angry that residents in the community across the road were relocated while they were left behind, blamed CATE and Margaret Williams for having failed to advocate for them forcefully enough. But Wade does not join in this castigation. She says that Williams opened the door for them on the relocation issue, but they will have to fight to walk through it on their own. To this end, Wade makes a strong case that residents cannot stay where they are. "We have no doubt that the pollution is causing health problems including cancer, birth defects, premature births, and autism," she says. Soil testing performed by EPA officials found a maximum of 85.3 ppt of dioxin in the soil compared with an allowable residential standard of 3 ppt.[35]

The EPA wanted to resume excavating the Superfund site, but Wade insisted that Clarinda Triangle residents need to be moved out of harm's way before the digging starts. After numerous meetings with regulatory officials and writing "fierce" letters to the EPA, an agreement has been reached for relocation, she says. The EPA estimates that it will cost $3.15 million to clean up and relocate residents in Clarinda Triangle if it is cleaned up to a level consistent with future commercial use of the area. The estimate assumes that fifty-five households will need to be relocated, at an average cost of $57,000 per home.[36] The first six families were relocated on November 20, 2006. An updated EPA plan to move an additional ten homes estimates the total cost per home at $83,273, a sum that includes the cost of the replacement home, the moving expenses, and the demolition of the old home.[37] "This is not adequate to cover the replacement cost of comparable houses and moving expenses," argues Wilma Subra, a chemist and technical advisor who advised resident groups.

**Stolen Dreams**

When Catherine L. Clark, age fifty-nine, moved to a double-wide trailer in Clarinda Triangle in 2000, she had a dream. She wanted to establish a church with a garden attached to it where she could grow fruit trees and vegetables to feed her neighbors. The garden would also have a playground for local children. For years she invested in the garden, worked the soil, and planted trees. Then men in hazmat suits informed her that the soil was contaminated and not to eat food grown in it. "They took away some of our dreams. I'm saddened by that," she says.

Clark is a big woman with a big family. She has five natural sons, three stepsons, and three adopted sons. Five of her other children died in childbirth. She also claims sixty-four grandchildren and twenty-two great-grandchildren. To be in her house is to be at the heart of one of the kitchens that feeds a good part of the neighborhood. Deep pans filled with hot food are either in the oven or awaiting transport on the counter covered in aluminum foil. Clark distributes the food to some of the community's elderly, ill, and bedridden residents who can no longer cook for themselves. Since the tap water is polluted, Clark has to buy bottled water for drinking and cooking. "I spend hundreds of dollars on water," she laments.

Clark took her ministry to Escambia Arms, the subsidized housing complex across from the wood treatment plant. There she found that the children who played in the dirt had eczema so severe that their arms were held out stiff from their bodies. One child had big boils under his arms. Her daughter-in-law lived there and had a child with pinholes in its heart. Clark's garden became an oasis for the children and provided what she thought was healthy organic food for many. She grew okra, corn, eggplant, squash, melons, yams, potatoes, herbs, tomatoes, onions, and garlic and planted fig and peach trees. "That was my life. That's what I lived off," she says. Her plan was to create a family homestead with four separate homes for herself and her sons. "Now it's been ruined," she says.

"The government allows these businesses to set up and they don't check to see if they are detrimental. They should have had regulations in place and not allowed that [heavily polluting industries] in this place," she continues. "I decided not to build a day care center here because of the contamination. I saw all the kids with skin problems and tried to take care of them. They had sores all over their bodies. They cried, and bled, and scratched. Many of the kids had birth defects or mental retardation. They won't be able to take care of themselves and will become a burden to society."

The people who caused the pollution should be held responsible for the harm they did, Clark maintains. The corporations should step up to the plate and pay: "They left the burden on the little man's shoulders," she opines. They should also pay for follow-up monitoring of the health of the people who have been exposed, she continues. Clark is in favor of an expedited relocation program for home owners but is also concerned that renters should be fairly treated. Renters were exposed to contaminants, and many of them were never given health screenings, she notes. "The Clarinda Triangle Association is about more than relocation. We also want to make sure that no one is left behind. "The young and the elderly should be moved out first. I won't budge until everyone has been moved. . . . I can be the last."

**Reburying Mount Dioxin**

What to do with Mount Dioxin remains a quandary and source of debate among federal, state, and county officials, as well as city residents. Many agree that something must be done given that the plastic covering the highly toxic wastes is wearing out. But just what kind of remediation should take place and whether the wastes should be treated, removed, or reburied on site remain contentious issues.

County and federal officials are anxious to finish the remedy at Mount Dioxin and turn it into a commercial park. County officials declared the area a Community Redevelopment Area in 1995, and the EPA designated it a Brownfield Pilot Program in 1998, providing some funds for cleanup.[38] But what technical solution should be used to treat or isolate the wastes remains in question. The EPA's preferred remedy is to rebury the wastes on site along with contaminated soils from the surrounding communities in an expanded pit. After CATE objected strenuously that the contaminated wastes should not be dumped into an unlined pit, the EPA agreed to line it with heavy plastic, Subra notes. The EPA then proposes to cover this new plateau of reburied wastes with one layer of solidified wastes (comprising 13.5 percent of the total), a layer of plastic, a layer of clay, and then top it off with native soils and grasses. Short of doing nothing, which is what has happened for the last fifteen years, the so-called capping/containment alternative is the cheapest remedy available, estimated to cost $24.3 million.[39]

Just burying the waste on site, however, is not as safe as treating it or removing it, observes Subra. "The contaminated soil will not be treated to reduce toxicity," she explains, and the buried soils will remain a potential source of contamination. The plastic liner that authorities plan to use has only a thirty-year expected lifetime and is not nearly as sturdy as the ones used to line hazardous waste disposal sites. "Holes and flaws in the geomembrane liners will allow wastes to leak into the groundwater," Subra predicts.[40]

The depth at which engineers plan to rebury the waste is particularly problematic on this site because it is only five feet above the high groundwater elevation, Subra continues. In other words, this large volume of untreated, highly toxic wastes will be separated from the high groundwater mark in the shallow sand and gravel aquifer by only a leaky piece of plastic and five feet of soil. This likely will lead to further pollution of the groundwater, which is already heavily contaminated, and, she concludes, is "not an acceptable remedy." Other solutions, which involve either treatment or removal of the contaminated soils, are expensive: thermal desorbtion treatment would cost $246.7 million, more than ten times that of capping it. Bioremediation treatment would cost $157.9 million, and transporting the waste off-site to a licensed hazardous waste dump would cost $312.8 million, a sum the government is unlikely to pay given the reduced financial resources available for remediating Superfund sites. Finally, in the soil solidification and stabilization option, wastes would be stabilized with a binding agent such as concrete and then reburied. This option, which would cost an estimated $51.8 million, is also no bargain because binding the contaminated soils together is not the equivalent of treating them, so they will remain hazardous, Subra points out. The suggestion by EPA officials that they might spend an additional $2.3 million solidifying and stabilizing some of the most toxic wastes will affect only 13.5 percent of the material, she adds.

Reburying the wastes without treating them is not only ill advised from a public health standpoint; it is also against Florida law: "The state of Florida has a prohibition on landfills for such waste," Subra notes. Nevertheless, a deal has been made to go ahead with the reinternment of Mount Dioxin. "The issue of redevelopment [the Commercial Park initiative] drove the remedy rather than environmental and public health protection," Subra concludes.

**Argument over Remedies**

By June 2007 the EPA was poised to begin work on what agency officials see as the final phase of the remediation effort.[41] The job entails encasing over a half-million cubic yards of soil on site at a cost of some $17 million and will take some sixteen months. In all it is expected that some four hundred residents of neighborhoods surrounding the site will be relocated. While Williams, Dunham, Subra, and other Pensacola residents downhill from the site contend that burying the wastes on site is not the best solution. EPA officials disagree: "The EPA still feels this remedy is going to be protective of human health and the environment," a local reporter was told by Laura Niles, spokeswoman for the EPA's Southeastern Regional Office in Atlanta.[42] But excavating and moving 400,000 cubic yards of waste is going to stir up a lot of contaminated dust, and once again residents living in communities near the site will be in danger, notes Subra. Officials should consider relocating residents in neighborhoods across the railroad tracks from the site and at the nearby Brown Barge School, she adds.

**Carrying On**

Margaret Williams, now eighty years old, is ailing and doesn't have the energy to carry on the work of CATE without help. Her daughter, Francine Ishmael, forty-six, a former social worker, has taken over the day-to-day operation of the nonprofit that is now tracking and providing health screenings for the some six thousand former residents who lived near the neighborhood Superfund sites. To date eighteen hundred residents have participated in the program, a joint effort of CATE and the Escambia County Health Department [43]

Ishmael, who lived at her mother's old house on Pearl Street for seventeen years, returned home in 1993 and joined CATE the year after it was founded. She watched her mother wrestle with regulators over relocation and learned a lot about the extent of the contamination she had grown up amid. It prompted her to reflect on the possible causes of ailments that she and other residents had suffered. "There were a lot of unexplained rashes among kids, headaches, nosebleeds, and behavioral problems. The ladies in my class at school later had too many children with birth defects," Ismael observes. One of them sticks in her mind: a child who was born with one disproportionately large arm. Ishmael's mother also has a granddaughter born with six toes, an abnormality that was later corrected through surgery. A lot of the young women Ismael grew up with had reproductive problems and hysterectomies.

This cluster of cancer, birth defects, and reproductive disorders caused Ishmael to wonder if her illnesses had been caused by exposure to pollution from the plants. "I drank the water and played in the dirt near these places," she recalls. "I was clueless at the time. It never dawned on me that playing in the dirt or riding my bike might cause me harm despite the fact that I had asthma." After leaving for college in Gainesville, she started having medical problems. "My stomach suddenly swelled up as if I were eight months pregnant," she recalls. When she went to the hospital, she was immediately operated on for a large abdominal abscess that required both small bowel surgery and a hysterectomy. At the age of twenty-three, she spent six weeks in the hospital recovering. "I almost died," says Ishmael who learned from her doctors that she would never bear children.

The pattern in her family was too clear to ignore. One grandmother had breast, ovarian, and colon cancer; her other grandmother had four miscarriages; her mother had two children born either dead or so sick they died within three months; and now she was the third generation experiencing major reproductive problems. "I began to think about all the things that might have contributed to my problem," Ishmael says. She remembered gardening and digging in the soil, eating the delicious plums, grapes, and blackberry cobblers they made from the fruit they grew.

"The government did not serve us well," concludes Ishmael. Residents were not notified that they were living next to two heavily polluted sites and that a massive cleanup was about to take place. The neighborhood she grew up in was so polluted that when it rained, an oily sludge bubbled up out of the ground. "It is really awful. The government does not have the courtesy to think about our health. We have six Superfund sites in our county but in the affluent neighborhoods, there are no industries or treatment plants," she points out. Despite these challenges, Ishmael learned that "people have more power than they think they do if they work together."

### Minimizing the Problem

Federal, state, and local officials downplayed the possibility that contaminants from the plants had caused health problems among nearby residents, notes Frances Dunham of CATE. Regulators said there was an odor problem but not a health problem and claimed what they had was "a soil emergency," not a health emergency, she continues. Both an official from the Agency for Toxic Substances Disease Registry (ATSDR) and a state health official said that it was impossible to tell if pollutants had caused ill health in the community and suggested instead that most of the illness was caused by lifestyle issues such as smoking, drinking, and drugs.[44]

Conditions surrounding Mount Dioxin do "not represent an emergency situation," said Mark Fite, the EPA Region IV project manager in 1996, noting that contamination levels were not so high as to pose threats to residents. "We're within the risk range," he added.[45] Activists think these statements were misleading. Federal environmental and health regulators "provided false reassurance to officials and physicians: they treated victims [of environmental contamination] as neurotics if not opportunists," Dunham charges.[46]

At the federal level, the ATSDR raised no health concerns about dioxin despite the fact that soil samples had found readings as high as 950 ppt, while the standard residential exposure level was 7 ppt. ATSDR officials took the position that "we do not know what cancer health risks are likely" due to the fact that dioxin's carcinogenic properties are under review.[47] This was hardly the precautionary principle in action.

Despite the fact that the ATSDR did not raise a health alarm over pollution problems around Mount Dioxin, independent experts point to a striking similarity between the type of diseases that residents suffered and the known illnesses caused by the chemicals to which they were being exposed. For example, dioxins and furans, which can be lethal at minute doses, were found at very high levels on the grounds of the plant and in the soils in surrounding neighborhoods. Dioxin is a human carcinogen, and furans are possible carcinogens. They also cause nervous system disorders, skin disease, liver and kidney disease, birth defects, and reproductive disorders, along with a host of mental and behavioral problems. All of these problems sound familiar to residents in the neighborhoods surrounding the plants.

Elevated levels of dieldrin, a banned pesticide, were also found in the soil of homes near both ETC and the Agrico Chemical Company. Dieldrin was banned because it is probably a carcinogen and is known to cause liver, kidney, nervous system, and reproductive disorders. Levels of dieldrin as high as 2,000 ppb were found in the soil outside homes near the chemical fertilizer factory; a level of 40 ppb is considered safe.[48]

Near ETC, levels of arsenic in the soil were found as high as 9,400 ppb, and outside the grounds of the Agrico Chemical Company, they reached as high as 48,000 ppb. Both readings are considered high, given that the safe level is 370 ppb. Arsenic can cause bladder, kidney, lung, and liver cancer, as well as irritation to the skin, eyes, and throat, and abnormal heart and nerve function.

Finally, 1,133 ppb and 24,705 ppb of benzo(a)pyrenes were found in the soil outside the fencelines of ETC and Agrico, respectively. The EPA considers 88 ppb a safe level. These chemicals are known to cause lung cancer, leukemia, tumors of the stomach and skin, and developmental and reproductive effects.[49] Mix all of these chemicals together into a toxic brew, and then expose people of all ages to them over a period of decades, and some serious medical problems are likely to result.

## The Golden Rule

With constrained budgets for the cleanup of Superfund sites, regulators and engineers face stark ethical dilemmas. How many people should be moved away from these sites? Should residents be told of dangers when remedial excavations begin? These moral quandaries are now occurring with such frequency that workshops are being held on the subject to train engineers to make decisions critical to the life and well-being of local residents exposed to industrial toxins.

At one such workshop, where the ETC excavation was used as a case study, presenters suggested that the Golden Rule might be the best guide for engineers faced with the cleanup of a Superfund site adjacent to a residential community. Site managers responsible for the cleanup operations should ask themselves if they would be "willing to exchange places with those living next to the site," write Edmund Tsang and John C. Reis. These authors very much doubt that site managers would have been willing to move into neighborhoods adjacent to Mount Dioxin knowing that the toxins involved caused cancer and birth defects. Not immediately informing residents of the health dangers they faced from chemical exposures also violated their "right to free and informed consent and possibly the rights to life and health as well," they conclude.[50]

Applying the golden rule is almost always a good place to start. But as Lois Gibbs points out, invoking high principles and telling the truth about contamination in residential neighborhoods is not enough to cause change. Often protracted, tenacious grassroots campaigns backed with assistance from national environmental justice groups are needed. This was the case in Pensacola.

## Few Outright Victories

Most successful grassroots environmental justice campaigns are spearheaded by strong local community activists who take up the cause of protecting their family and neighbors from harmful chemical exposure. The neighborhoods on the fenceline with the Superfund sites in Pensacola were fortunate in this regard. Guided by the leadership of Margaret Williams, the CATE coalition of neighborhood activists won a package of relocation benefits after years of hard work. "Persistence, unity, and visibility have paid off," writes Williams, looking back on the long struggle. Convincing EPA officials to relocate the neighborhood was "an important victory for local residents and also, we hope, for the principles of public health protection," she concludes.[51]

The "victories" won by environmental justice activists are often qualified victories that do not work equally well for everyone. In Pensacola, for example, many of the payouts that residents received to relocate and purchase replacement homes were miserly when compared with compensation allocations given to white residents at other sites. But at least the beleaguered residents near Mount Dioxin in Pensacola gained an opportunity to sell their properties and move out.

Seen in the context of the thousands of other residential communities that continue to be exposed to disproportionately high levels of industrial chemicals from neighboring facilities, the environmental justice victories—such as those scored in Ocala and Pensacola, Florida, and in Norco, Louisiana—are relatively rare. One reason is that in many sacrifice zones, the source of pollution, such as a giant refinery, is too large and powerful to shut down; or the adjacent residential area that is affected is so large that it is difficult to convince corporate, state, and federal officials to pay for the relocation of large numbers of people. In many of these cases, the best deal available may be a reduction in emissions and financial help for citizen to purchase air monitoring equipment and health services for pollution-afflicted residents.

In the next chapter we look at a low-income African American neighborhood in Port Arthur, Texas, on the fenceline with one of the largest refineries in the nation, where neighboring residents suffered from high asthma, respiratory disease, and cancer rates. The grassroots group that coalesced in this area, capably led by Hilton Kelley, eventually won a promise of significant emission reductions and investments in pollution control equipment. They did so by leveraging the refineries' desire to win a permit to expand operations in return for reduced emissions and financial help for the nearby community. To date, however, residents have not convinced local polluters to pay for the relocation of the most vulnerable residents, some of whom live in public housing immediately adjacent to the refinery.

# II
## Contaminated Air

# 3

## Port Arthur, Texas: Public Housing Residents Breathe Contaminated Air from Nearby Refineries and Chemical Plants

Hilton Kelley is a big man, forty-five years old, with a shaved head and a brown belt in tae kwon do. He grew up on the front lines of toxic chemical exposure in the United States in the West Side neighborhood of Port Arthur, Texas, where he lived in the Carver Terrace subsidized housing apartment complex just across the fence line from a giant refinery. The Motiva Enterprises refinery, covering thirty-eight thousand acres, remains the eight-hundred-pound gorilla on his block producing 285,000 barrels of oil a day. Refinery officials plan to expand capacity 125 percent and produce 625,000 barrels a day, making it the largest refinery in the nation.

This is a tough neighborhood when it comes to air pollutants. The Motiva facility, jointly owned by Shell Oil and Saudi Aramco, released over 15.5 million pounds of criteria pollutants in one year, ranking in the top 10 percent of the dirtiest plants in the United States.[1] Motiva emitted 14.9 million pounds of criteria air pollutants during routine operations in 2003 and another 648,400 pounds during emission events and maintenance, start-up, and shutdown activities.[2]

Sometimes the refinery had as many as four unscheduled releases of multiple chemicals on a single day.[3] On April 14, 2003, Motiva emitted 274,438 pounds of air contaminants: 107,280 pounds of hexane, 24,607 pounds of butane, 29,424 pounds of heptane, 11,834 pounds of isobutene, 37,538 pounds of pentane (toxic to the central nervous system and causing fatigue, irritability and other behavioral changes), and 14,992 pounds of propylene (toxic to the respiratory system). Many of these pollutants can also create ground-level ozone pollution (smog) that causes breathing problems and aggravates asthma.[4] On April 15, 2003, the plant emitted about 9 tons of particulate matter while children were waiting at bus stops on their way to school. Some fifteen residents called regulators to complain of heavy black smoke, bad odors, and soot falling on cars, and others complained of health problems such as headaches and children with asthma.

Heavy emissions drift into the neighborhood from the Valero refinery, Huntsman Petrochemical, and the Chevron Phillips plant, as well as the Great Lakes Carbon Corporation's petroleum coke-handling facility. The air is further burdened by massive releases from a major refinery owned by Total Petrochemicals USA (formerly Final Oil), Premcor Refining, and BASF Fina Petrochemicals located a few miles away in East Port Arthur. One mile to the west is a hazardous waste incinerator owned and operated first by Chemical Waste Management and subsequently by Onyx Environmental Services, which pumped out over 100 tons in criteria air emissions in 2004.[5] Some of these air pollutants find their way into the lungs of residents who live in adjacent residential neighborhoods.

### "Why Is That Kid Barking?"

In the sweat-popping heat and humidity of summer in Port Arthur, Edward Brooks II, an unemployed heavy-equipment operator, stands next to his Carver Terrace apartment dressed in sleeveless T-shirt, checkered pajama bottoms, and slippers. Inside his apartment it's a shade cooler with the air conditioner on full blast. On many days, however, fifty-six-year-old Brooks has to switch off the air conditioning to prevent it from sucking toxic fumes into his home from the neighboring refinery.

The air wafting out of the refinery and over the fenceline into the government-subsidized housing where Brooks lives is sometimes tolerable and sometimes flat out awful, he reports. "Half the kids here need help breathing, and a lot of them have asthma medicine they take to school and breathing machines at home. You don't need to be a rocket scientist to see that this isn't normal," he adds. "You hear these kids gasping for air and someone will say: 'Why is that kid barking again?' Some of these kids can't run half a block their lungs are so bad." Brooks's wife is also afflicted by the contaminated air and suffers from such severe respiratory distress that she must depend on an oxygen tank situated close to her bed. "We're trying to get out of here, but we can't afford to move," Brooks explains.

The association between heavily polluting point sources of emissions and elevated levels of asthma is well documented. A Harvard School of Public Health study in 2000 "linked 43,000 asthma attacks and 300,000 daily incidents of upper respiratory symptoms per year" and 159 premature deaths a year to pollution emitted by two giant power plants in New England, eastern New York, and New Jersey.[6] This makes Brooks's contention of a link between pollution coming from nearby plants and asthma and respiratory distress in his neighborhood highly credible.

"This area is not safe. We are three hundred or four hundred yards from the refinery here. I want to get my family away. We want to move so we can get a chance to live," he says bluntly. But coming up with the money to move is a problem in a community where the official unemployment rate is 13.5 percent and the actual rate of unemployment is much higher. "Anyone with any knowledge knows they should move on. The government is not doing anything to protect us. They tell us about the emissions, but they don't do anything about it," he charges. "If I had any power, no one would be allowed to live here. This place should be crushed to the ground," but most people can't afford to move out, he explains. They have subsidized apartments on the fenceline in public housing complexes like Carver Terrace, Lewis Manor, or Prince Hall and don't dare give up their apartment for fear that they will be out on the street, he adds.

## Dangerously Sited Public Housing

Hilton Kelley agrees with Brooks: "Some of the folks who breathe this air too long die of cancer. As we speak, kids are being born who are being brought back to these projects to breathe in toxic air. That just isn't right. We need to clean up this place up for the new souls," he says. Kelley has personal experience to back up these words. He grew up within sight of Motiva at Carver Terrace and breathed air laced with elevated levels of benzene, sulfur dioxide, hydrogen sulfide, and 1,3-butadiene. The air smelled like rotten eggs from the sulfur, he recalls. "At night we had a bright orange sky because the refineries were constantly flaring, burning off fumes and gas."[7]

Despite this breath-taking pollution and the few comforts of public housing, Kelley describes his early days with some fondness. There was food on the table, and his mother kept Kelley and his younger brother busy so they wouldn't get in trouble in the streets. The boys attended karate classes, YMCA, Boy Scout meetings, football games, marching band, and church. "I was always in some kind of uniform," he recalls.

But at the age of eighteen, this tightly scheduled routine ended abruptly: on February 27, 1979, Kelley's mother was shot to death. A year later, he joined the Navy, trained to be an electrical engineer, and served on the *USS Roanoke Relay*, an oiler that shuttled jet fuel to aircraft carriers. Leaving the Navy in 1984, Kelley landed a series of acting jobs, first in Hollywood and then in northern California, including a part on the *Nash Bridges* cop show with Don Johnson.

With his Screen Actors Guild card in his pocket and a TV show to his credit, Kelley came home an accomplished graduate of West Side Port Arthur's hard streets. It was on one of his visits home that he was struck by both the impoverishment of his old neighborhood and how bad the air quality had become. As a boy, Kelley took the pollution for granted, along with a constant cough and a skin rash, common in the neighborhood, that left him with little black spots all over his arms, chest, and back. Both problems disappeared once he left Port Arthur. "Everyone knew when the plant did a smelly, but no one did anything about it," he recalls. His grandfather, who lived five blocks from the fenceline, died of cancer. "Everyone had respiratory problems, sinus problems, skin problems, and allergies, but I thought that was the way life was," he says. But after traveling widely with the Navy, Kelley learned that heavily polluted air wasn't normal.

Port Arthur residents developed ways of coping with the foul air, which Kelley describes as periodically "so bad that it can take your breath away." Annie Edwards, a long-term resident, recounts the feeling she experiences when heavy gusts of pollution engulf her: "Like I panic and can't catch enough air and if I go outside it's worse. I have to strap on my breathing machine [oxygen supply] at night so I don't pass out."[8]

Smelling the bad air in his old neighborhood and seeing multiple families he knew with respiratory problems and other pollution-induced diseases affected Kelley deeply: "I couldn't get it out of my mind that I needed to do something for my hometown. Because of the increasing air pollution, the people of Port Arthur were too sick to help themselves.

They were beat down. The town was dying, and I saw a need that I thought I could fill."[9]

## "We've Been Waiting for Someone Like You"

Concerns among residents about air quality were nothing new in West Side Port Arthur. Throughout the 1990s, Reverend Alfred Dominic, a retired water company employee, stood up in city council meetings and talked bluntly about the health problems that residents suffered as a result of the pollution. Dominic, seventy-eight years old, is a well-known figure in the community. He is the father of thirteen children, and his living room walls are hung with photos of a number of his children who have served double tours in Iraq. A member of the Masons and Eastern Star Church, for years Dominic demanded that the city council take action to protect public health but was ignored. The reason, he says, is not hard to figure out: 80 percent of the city's tax base is paid by industry. "Industry is this city's bread and butter. Industry has influence in the schools, the churches, and the hospitals, so no one talks about the connection between high levels of pollution and the large numbers of kids with asthma. Most people here over forty-five know someone in the petrochemical industry," Dominic observes. "It's a company town."

When Kelley returned to Port Arthur in 2000 and showed an interest in redevelopment of the West Side, Dominic was relieved and ready to pass the campaign over to him: "We've been waiting for someone like you," he said. At the time Kelley was scrambling to make a living doing plumbing and electrical jobs and had no car to get to work. To help him get on his feet, Dominic gave him rides to his jobs and fed him in his kitchen. "He has been like a father to me," Kelley says fondly. Over meals they would talk strategy about how best to help the community. Kelley wanted to open a new community center, but Dominic questioned this approach and suggested that the area was too polluted to become a safe place for people. Instead he recommended fighting for funds to relocate the residents out of harm's way. "He educated me about all these issues and introduced me to the city council members," Kelley notes.

Sitting on his couch in his West Side living room surrounded by stacks of newspaper clippings and videotapes of himself testifying before the city council, Dominic retains his passion for protecting the health of local residents, but his health is clearly failing. A tank of oxygen is close at hand. Outside, a train carrying petrochemical products rumbles by, shaking the small wooden house and making conversation momentarily impossible. "We know that something is in the air and in the water that people used to drink," says Dominic, who was born on Fifth Street and has lived in the neighborhood most of his life. "All this pollution affected my health. I have problems with breathing, problems with nausea, and now I have problems with forgetting. Many of my friends have died of cancer, and many of them are sick at the present because of the emissions."[10]

### Reaching Out for Help

Grassroots campaigns protesting pollution are difficult to mount without help from the outside. Among those who gave Kelley assistance were Neil Carman, a former state inspector of refineries who turned whistle-blower and now is director of the Lone Star Chapter of the Sierra Club's Clean Air Program; Wilma Subra, the McArthur prize-winning chemist from Louisiana; then Refinery Reform director Denny Larson; Marvin Legator, the epidemiologist; and Eric Schaffer at the Environmental Integrity Project in Washington, D.C.

Kelley's introduction to the nationwide antitoxics movement came through Denny Larson, who now runs Global Community Monitor, a nonprofit that helps equip and train fenceline residents to conduct citizen-based air monitoring. In 1995 Larson trained residents in Port Arthur and Beaumont to conduct rudimentary air monitoring. Then in 2000, working through the Seed Coalition based in Austin, he orchestrated what he called a "toxic two-step" refinery pollution tour along the "chemical corridor" that runs through Houston, Beaumont, Port Arthur, Mossville, and New Orleans.

Larson also trained, equipped, and funded Reverend Roy Malveaux to do bucket brigade citizen air monitoring. A Baptist minister in Beaumont, Texas, Malveaux was preaching about the impact on his congregation of pollution from a large Exxon refinery located adjacent to an African American community on the south side of the city. Malveaux had made himself unpopular with oil industry executives in Beaumont and at his previous ministry in Corpus Christi, where he also preached against the perils of toxic contamination of his parish. For his trouble, he was ousted from both his churches by a dirty-tricks campaign run by the oil industry, Larson claims. "They hired the sons of deacons at his churches to work at the refinery and then threatened to fire them if they didn't get rid of Malveaux," Larson asserts. This infiltration tactic worked until Malveaux decided to start his own church.

Kelley attended these citizen air monitoring training sessions with Malveaux, and by early 2001, Larson helped him build an air monitoring device made out of a cheap plastic bucket and a Radio Shack air pump; paid for laboratory analysis of his air samples; and showed Kelley how to file a civil complaint and put on a press conference announcing the results, which showed elevated levels of butadiene and sulfur dioxide in the air. Before long Kelley started his own nonprofit, Community In-Power Development Association (CIDA) but he needed money to make it work. The Refinery Reform Campaign funded Kelley for three years until he had learned how to raise grant money. Through Larson, Kelley also met Peter Altman at the SEED Coalition, who took him to the Texas Commission on Environmental Quality (TCEQ). Along the way Kelley learned that "I had the right to look up information about accidental releases from the neighboring refineries," he recalls.

**Odor and Symptom Logs**

Kelley's antipollution campaign also built on previous work done in Port Arthur that demonstrated high levels of contaminants in the air and elevated levels of medical problems among local residents. In the early 1990s, before Kelley returned home, Wilma Subra held a series of workshops in Port Arthur. Her local contact on the West Side at the time was Luverda Batiste, who headed a group of residents concerned about pollution who called themselves Mentors Outlining Definitives for Earthly Living. Subra conducted a number of educational sessions in which she taught residents how to create an odor log that documented the date and time they smelled an odor, as well as a symptom log that chronicled any ill effects they experienced. She then collected these logs and matched them with publicly available Toxic Release Inventory information about emissions from nearby industrial facilities.

At first Subra's workshops were sparsely attended, but when she made it clear that the odor and symptom logs could be filled out anonymously, suddenly she had three hundred logs to analyze, and her workshops were filling a full-sized gymnasium. In the logs, residents reported odors they smelled and symptoms they experienced. "The data absolutely matched: the health impacts associated with the chemicals that were released corresponded to the health impacts experienced and recorded by the community," she explains.

Residents' notations in their odor logs indicating they smelled rotten eggs, sulfur gas, and chemical odors matched government reports of emissions of hydrogen sulfide, sulfur dioxide (the rotten egg smell) hydrocarbons, benzene, and ethyl benzene from nearby plants. During the same period, air samples collected by state officials and community monitors demonstrate that chemicals released by these facilities were crossing the fenceline into Port Arthur, Subra notes. Residents were exposed to increased levels of airborne toxins during industrial "upsets"— an unpermitted air emission that occurs when industrial machinery releases pollutants when it malfunctions or is turned off or on—on an average of five times a week. Some 75 percent of these upsets could have been avoided if refineries installed up-to-date pollution control equipment and new valves, she asserts.[11]

Using some of the data collected in the odor and symptom logs, Subra was able to convince U.S. Environmental Protection Agency officials to dispatch a mobile air monitoring unit known as a trace atmospheric gas analyzer (TAGA) truck to test the air in residential neighborhoods. Data collected by the TAGA truck from January 27 to 31, 2003, found six airborne chemicals in residential areas that exceeded the health effects screening levels. Among these were benzene (a known human cancer-causing chemical); chloroform (a possible human cancer causing chemical); vinyl chloride (a known human cancer-causing chemical); 1,2-dichloroethane (a possible human cancer-causing chemical) and 1,2-dibromoethane, which exceeded the one-hour, twenty-four-hour, and annual Texas Health Effects Screening levels; and 1,1,2-tetrachloroethane.

**Tracking Elevated Toxic Emissions**

With self-reported data from industry, data from the TAGA truck, and information collected through the odor and symptom logs, Subra presented maps to residents that showed where chemicals were released, which way the wind was blowing, where elevated levels of toxic chemicals were detected, and where residents reported odors and symptoms. When all this information was combined on a map, it made a compelling case that residents were being made ill by emissions from nearby industrial facilities. Community members were alarmed when they saw the maps, Subra remembers. They studied the map to see where they lived, worked, and went to church and then looked at the level of pollution reported. "Suddenly they were no longer afraid to ask questions and participate in events about how the health of their community was being impacted," she recalls.

With these chemicals crossing the fenceline, Subra describes the situation in Port Arthur as "dangerous" because residents are surrounded by heavily polluting industries that emit both permitted and accidental releases. Because the community is surrounded by industry, whichever way the wind blows, they are exposed to toxic chemicals. "The health of the community as well as the quality of their lives is impacted by the accidental releases," she continues. "The most frequently reported health impacts by community members were headaches, sore throats, burning or watering eyes, dizziness, or light-headedness, difficulty breathing or coughs." Other complaints were nausea, skin rashes, and nosebleeds. Subra also discovered a correlation between the time and date of accidental chemical releases and spikes in the use of emergency services by local residents.

## Health Impact

Texas toxicologist Marvin Legator, former professor of environmental toxicology at the University of Texas Medical Branch at Galveston, confirmed Subra's impression that pollution from nearby plants was damaging the health of Port Arthur residents. Legator conducted a health study in Port Arthur and another nearby community, Beaumont, also located near chemical industrial complexes. This study compared symptoms of adverse health effects in Port Arthur and Beaumont residents with a control population of residents in Galveston. What he found confirmed what fenceline residents already suspected: their health was being compromised by exposure to high levels of pollutants. "Without question, the people in Beaumont and Port Arthur are suffering from many more health problems, especially neurological and respiratory diseases, than those in Galveston. The concentration of heavy industry there [in Port Arthur] is having an enormous impact on their lives, and this study proves that to be the case," Legator told a reporter from the *Texas Observer*.[12]

In his study, Legator, who died in 2006, found that approximately 80 percent of the residents he interviewed in West Port Arthur reported cardiovascular and respiratory problems compared with much lower levels (approximately 30 percent for cardiovascular and 10 percent for respiratory problems) in an economically and racially similar community in Galveston. Similarly, approximately 80 percent of residents on the fenceline suffered from ear, nose, and throat problems compared with approximately 20 percent in the control group, and almost 75 percent of West Side residents had complaints such as headaches compared with approximately 30 percent of those who lived in a less polluted area. Another study found that school absences among the 21,800 children who went to school within two miles of a petrochemical plant in Jefferson County also increased following large accidental releases of toxic chemicals.

**Getting Organized**

To protest the vast volume of toxic emissions invading West Side Port Arthur that was causing illness among residents, Hilton Kelley educated himself in the arcane language of toxicology and epidemiology so that he could make a case that his community was being poisoned. Kelley launched his campaign for cleaner air by standing on a corner with a sign accusing local refineries of flaring off toxic chemicals illegally and the regulatory agencies of ignoring the problem. During flaring off operations plant operators attempt to burn off chemicals at the top of smokestacks that would otherwise be released into the air. Frequently, this burning off of by-products is incomplete, and large quantities of contaminants are released near residential areas.

Kelley walked the neighborhood knocking on doors and passing out flyers. As he talked with residents, he did his own informal health survey and found that one in five households had a child with asthma who required medication or breathing treatments; others had a distinctive type of skin rash associated with pollution. Kelley told residents that it was not their fault that their child had asthma and that the regulatory agencies should be doing more to reduce pollution. "We are trying to push industry to clean up emissions and use up-to-date [pollution prevention] technology," he says.[13]

He also argued that the refineries should pay for medication for the diseases such as asthma that their emissions were either causing or aggravating. Kelly discovered evidence that asthma was more prevalent in Jefferson County where Port Arthur is located than in other parts of the state. In Jefferson County, 7.14 residents per 1,000 suffer from asthma compared with 5.52 per 1,000 across the state. Similarly, Jefferson County residents on the fenceline with heavy industry had 393.6 hospital admissions for chronic obstructive pulmonary disease per 100,000 compared with a state rate of 215.4 per 100,000.[14] These data suggest that West Side residents were literally chocking on air pollution being put out by local industries.

Kelley also told local residents that the petrochemical plants are fined less than 1 percent of the time when there are accidental releases and upsets and that when fines are imposed, the money goes into the Texas state coffers instead of being returned to the affected communities. As a result, he notes, residents afflicted by the pollution are not compensated for the harms they suffer.

### Encountering Resistance

Kelley's early efforts to stir up community interest in protesting the elevated levels of pollution and disease in West Side Port Arthur met with little success. The first official meeting of CIDA was attended by only two people: Kelley himself and the person in charge of the meeting room. Organizing residents to protest pollution was not easy because most of them struggled to pay the rent and keep the lights and gas on, leaving them little time and energy to protest pollution from neighboring plants, he explains.

Hush money also played a role and was used by heavy industry to mute criticism of their emissions, Kelley charges. He was once offered $4,000 and four computers by refinery officials for CIDA, but the offer evaporated after he made it clear that if he accepted the gift, he would in no way mute his protests about pollution. Local churches, civic groups, and politicians who depended on the largess of the big petrochemical companies also are hesitant to speak out against pollution. One community leader, who raised corporate money for local scholarships, told Kelley if he kept complaining about pollution, corporate donations might disappear. Cash settlements for pollution claims also dampened criticism of industry. Sometimes following a large chemical release, plant officials walked around the neighborhoods offering $50 in cash to residents willing to sign a document saying their complaint had been satisfied.

Despite these obstacles West Side residents gradually listened to Kelley's message and understood the need to speak up if their health complaints were to be heard. As a result, CIDA now has a membership of 120 members and a core group of 30 who can be relied on to show up at protests and news conferences. Although these numbers are still small, Kelley has effectively reached out to other environmental justice activists, academics, and foundations to make the case nationally that Port Arthur residents should not be exposed to the high level of toxic releases that are routine in their neighborhood.

As he began to "crusade to empower citizens to fight for their health," Kelley started to view his neighborhood as a sacrifice zone: "Our neighborhood pays the price for the rest of the nation's cheap gas," he observes. The equation is simple: refineries minimize their investments in pollution control equipment and as a result can raise profits and keep gas prices lower than they would be if they operated in a way that protected the health of their neighbors.

**Residents Demand Reduced Emissions**

In early July 2006, Kelley sat alone in his office, housed in one room of a single-story, white-brick storefront next door to Cash Loans & Anything of Value Pawn Shop. He was slogging through a telephone health survey of West Side residents. "Do you or anyone in your family have cancer?" he asked a local resident who answered his call. About 35 to 40 percent of households on the West Side have someone in their family who has died of cancer, Kelley estimates. The incidence of women from ages fourteen to fifty who have fibroid tumors in their uterus is also elevated, he adds.

"We are not trying to shut down these petrochemical plants," Kelley explains. The refineries provide thousands of jobs in Port Arthur and are an important part of the local economy: "We just want them to clean up their act." Residents shouldn't have to choose between working in an unhealthy environment and putting food on the table, he explains.

As part of his organizing outreach campaign, Kelley makes frequent visits to the stacked living units at Carver Terrace where he once lived. When I visited in 2007, the public housing complex had recently been damaged by hurricane-force winds and was patched with huge blue tarpaulins tied over the roofs. At one corner of the complex, near a heavily polluted area where fuel storage tanks had once stood, the belongings of an evicted tenant lay on wet ground exposed to the elements. With the huge Shell/Motiva works looming behind them, a couple of people poked through the apparently abandoned personal property to see if there was anything worth salvaging.

Dressed in a bright yellow T-shirt and cap emblazoned with CIDA's motto, "A United Voice for and by the Community," Kelley toured the complex like a mayor visiting his constituency. He talked with residents about the local pollution problems and urged them to join in news conferences and protests designed to raise awareness about the contamination. Many knew him, and some came to him with their problems. When a young woman who was being evicted for failure to pay her utility bill approached him, Kelley came up with two hundred dollars to keep her from being put out on the street in return for a promise to get her life in order and plan her finances more carefully. He also urged her to attend an upcoming CIDA meeting.

Some residents gave eyewitness accounts of recent releases. "There was some kind of green smoke that came out of the plant last week. Then the cloud turned to orange," reports Laura Paul, pointing toward the Motiva refinery that stretches out across the street from her home. For the past four years, Paul has lived in an apartment in the two-story orange-brick buildings at Carver Terrace, a housing complex of 384 apartments, subsidized by the Department of Housing and Urban Development, which is laid out in sixteen rows of twenty-four identical multi-housing units. Paul has a ten-month-old baby with bronchitis, and her mother was recently taken to the hospital for emergency treatment of a respiratory problem. "She couldn't catch her breath," Paul explains. "We are closed in by refineries and pollution here, and it is affecting the whole community."

## West Side Port Arthur

The West Side neighborhood of Port Arthur, located across the railroad tracks from the more affluent part of the city, was never a high-rent district, but it once was a lively port that sailors visited when their ships docked. It was not uncommon to hear Russian, Spanish, Arabic, and a host of other languages spoken in the streets when sailors were on leave, Hilton Kelley recalls. There were nightclubs, pool halls, bars, and brothels in one part of the neighborhood; elsewhere there were quieter working-class residential areas with grocery and ice cream stores amid the residential homes. Since the refineries expanded their operations and pollution levels increased, residents who could afford to moved out, leaving a largely elderly and poor population behind.

Warren Kelley, Hilton's younger brother, is one of a number of local activists attempting to improve conditions on the West Side. He opened the Black Tiger CIDA Karate School, located a short drive from the Carver Terrace apartments where he and his brother grew up. Frighteningly fit, Warren Kelley became a karate champion at fourteen, a black belt at seventeen, and won a title at twenty-one. He opened the karate school to give kids a place to go after school, and twenty-five signed up. The kids come to the dojo, show Kelley their school grades, and then begin to practice. Linda Simpson, his partner, says that many of the children have chronic asthma, bronchitis, and sinus problems because of pollution from the neighboring industries. "You eat right and still this happens to you," she says. "It gives you the feeling that you are being violated."

"In this neighborhood you have fireworks when it is not the Fourth of July," Warren Kelley observes, referring to accidents and flares that light up the sky over the refinery. The Kelley brothers are not the only ones who are struck by this eerie phenomenon. In fact, there are so many flares from heavy industry concentrated around Port Arthur, Beaumont, Port Neches, and Groves that the area is known as the "golden triangle" because that is what it looks like from an airplane at night.

"Many of the people I knew when I was growing up, if they didn't get killed in a car or by a gun got killed by cancer," Warren Kelley claims. Growing up with pollution from the refinery caused a wide variety of respiratory and skin problems that were not normal, as well as sinus pain and an impaired immune system, he continues. Recently, while repairing his roof, he was hit by a cloud of benzene from the plant that forced him to take shelter inside.

## Large Volume of Toxic Chemical Releases

Is there a connection between the high rates of respiratory disease, child-hood asthma, cancer, and skin rashes in West Port Arthur and the large volumes of toxic chemicals released from petrochemical plants next door? Common sense, the sheer volume of chemical releases, and the fact that many of the emissions are known to cause respiratory problems and cancer suggest that there is.

One way to get an idea about how dirty the air is on the West Side of Port Arthur is to examine the everyday "permitted" releases of toxic chemicals from surrounding chemical and petrochemical facilities and then add to that the toxics released into the neighborhood by "acciden-tal" upsets, flares, and start-up and shut-down releases. An accounting of these toxic emissions that rain down on the West Side was published by Eric V. Schaeffer, director of the Environmental Integrity Project (EIP). Schaeffer is a former EPA official who quit his job in a protest over the inadequate enforcement of the Clean Air Act. With his colleague Huma Ahmed, Schaeffer published two reports about chemical pollution in Port Arthur. The first, "Smoking Guns," details the huge volumes of chemi-cals released by flares; the second, "Accidents Will Happen," documents the releases of large volumes of toxic chemicals in a series of allegedly uncontrollable and unforeseeable accidental "upsets" which, on closer examination appear to be remarkably predictable. The scale of some of these accidents is staggering. The EIP report calculates that in the first seven months of 2002, heavy industrial facilities surrounding Port Arthur released almost 725 tons (1.1 million pounds) of toxics into the air of the city's fifty-eight thousand residents.[15] All on its own, the Premcor (now Valero) refinery near Port Arthur released 208 tons (416,492 pounds) of sulfur dioxide, as well as 25 tons of volatile organic chemicals (VOCs) and 2 tons of hydrogen sulfide in January 2002. The Motiva refinery, adjacent to Carver Terrace, emitted 4 tons of sulfur dioxide on April 7, 2002. Sulfur dioxide can cause burning in the nose and throat, breathing difficulties, changes in lung function, and asthma. It can be particularly hard on children, who breathe more air for their body weight than adults, the EIP report explains.[16]

Are so many accidents that release tons of hazardous chemicals into the air adjacent to residential areas really accidental? The authors of the EIP report suggest that while some releases are unpredictable accidents beyond the control of facility operators, most others are not and could have been predicted or avoided if proper maintenance, pollution control devices, and backup systems were in place. The problem is really systemic, the report's authors contend.

"The EPA needs to investigate the pattern of 'malfunctions' in Port Arthur, and take enforcement action to require better equipment or maintenance programs to eliminate pollution from accidents," Schaeffer argues. The laws also need to be tightened so that companies cannot dump tons of toxics into the air and get away with it by claiming it was an accident, he continues. "Polluters are expected to pay when their accidents release oil or chemicals into our water. We ought to demand the same accountability when the same chemicals are released into the air, where they may even be more threatening to the public's health, and degrade the quality of life in towns like Port Arthur," he concludes.[17]

### Elevated Cancer Risk

Industrial works such as Motiva and Valero and a host of other high-emission plants combine to generate enough pollutants to make Jefferson County, in which Port Arthur is located, one of the dirtiest counties in the country in terms of its air pollution load. It also ranks in the worst percentile for total environmental releases for increased cancer and other noncancer health risks.[18] The added cancer risk from hazardous air pollutants is higher in Jefferson County (670 per million) compared with the overall rate in Texas (550 per million).[19]

The release of large volumes of benzene in the area surrounding Port Arthur is of particular concern. Toxic Release Inventory data going back to 1997 show that 342,850 pounds of benzene were released in Port Arthur, 115,574 pounds in Beaumont, 54,666 pounds in East Port Arthur, and 1,829 pounds in Port Neches. These locations are all relatively close together, and toxic emissions float around the area and mix, explains Neil Carman of the Sierra Club.[20] The added cancer risk to Jefferson

County residents from exposure solely attributed to benzene is 54 cases per million compared with the state burden of 35 additional cancers per million.[21] Benzene is also a cause of leukemia, and Jefferson County males had a higher rate than the state for eight of the ten years between 1990 and 2000.

"This study shows a stunning failure of our state environmental regulatory agency, which has an obligation to the citizens of Texas to protect them from harmful air contaminants," the Public Citizen report observes. "TCEQ allows petrochemical companies to break the law, it does not impose penalties that deter violators, it allows companies to profit from harming public health, and it does not have an adequate monitoring system," the report states.[22]

**Toxic Trap**

In the face of this massive chemical exposure, Port Arthur residents have limited options. Most of those who could afford to move have already left; the majority of those who remain are unable to move for financial reasons. What is left to them is to organize, protest, and monitor the air. When a release is particularly intense, Kelley sometimes goes door-to-door warning people to get out of the area until the cloud lifts or shut their doors and windows and shelter in place. "When a cloud stays over our community for hours, you know it is a serious problem," Kelley observes.[23]

He also began to sample the air and captured readings of 4.2 ppb of benzene, which was substantially above the state long-term health screening level, the EPA regional screening level, and the Agency for Toxic Substances and Disease Registry intermediate minimal risk level. In addition, he found toluene, propane, and a host of other toxic chemicals in the sample he took. Once when Kelley was conducting this citizen air monitoring, officials from TCEQ arrived to investigate resident complaints but had none of the necessary equipment with them. "Are you here to watch?" Kelley asked. The irony of the situation was apparent: a local citizen was monitoring the air while the regulatory officials sent to investigate were empty handed. Significantly, despite the size of the release that day and Kelley's evidence of high levels of benzene in residential neighborhoods, TCEQ did not issue a violation against the company.

## Inadequate Regulatory Response

The struggle to improve air quality in Port Arthur and neighboring Beaumont has been a long-term campaign. In 2001, the Lone Star Chapter of the Sierra Club and other groups challenged the U.S. EPA's description of ground-level ozone (smog) problems in the area as "moderate." Arguing its case in the Fifth Circuit Court of Appeals, Sierra Club attorneys held that the smog problem should be reclassified as "serious." On April 29, 2004, the court agreed with the Sierra Club, required that the EPA elevate the air quality threat from moderate to serious, and moved up the date by which the air had to be cleaned up to meet the agency's one-hour standard for ground-level ozone from November 2007 to November 2005.

"The quality of air in Beaumont and Port Arthur communities has been unhealthy for many years, and we felt a new approach was needed that would accomplish improved air quality quickly," said Dr. Neil Carman, a former state inspector of refineries who turned whistle-blower and now is director of the Lone Star Chapter of the Sierra Club's Clean Air Program. Industrial polluters in the "golden triangle" around Port Arthur and Beaumont "were supposed to make large pollution cuts by 1996, but the EPA decided to give them until 2007 to clean up their act. Folks in this area shouldn't have to wait ten years for breathable air. So we felt we had no choice but to go to court," Carman said.

## Smog-Burned Lungs

"The Triangle [including Port Arthur] continues to have high ozone days each year when industrial pollution makes the air unhealthy to breathe. Ground-level ozone is a respiratory irritant, which has been shown to aggravate asthma, particularly in children. Adults with asthma and other lung conditions are also hard-hit by ozone. Breathing in too much ozone is like getting a sunburn on the lungs. It physically damages the tissue and kills lung cells," Carman explains.[24]

There was also an environmental justice issue: heavy pollution in the area falls disproportionately on minority and low-income residents, the Sierra Club and other environmental groups argued to EPA officials in a lengthy letter commenting on the agency's failure to enforce the Clean Air Act. Port Arthur's population is more than 50 percent poor and minority, rising to 100 percent in areas near the fenceline with industry, the letter continues. "Recent documents indicate that federal housing for poor minorities was allowed to be built in Port Arthur directly adjacent to these large polluting facilities with little regard for the health and welfare of those citizens," the report asserts.[25] "Beaumont-Port Arthur's minority and low-income populations are at elevated risk, are more susceptible to respiratory illness, are subject to higher concentrations of air pollution due to EPA's failure to administer the [Clean Air] Act," the letter adds.

## Deal to Cut Emissions

Following the lawsuit and court decision, a deal was cut whereby five petrochemical companies agreed to voluntarily reduce their emissions, and the state of Texas agreed to expedite a plan to reduce air pollution in the area to meet one-hour and eight-hour ozone standards. Polluting industries in the Port Arthur and Beaumont area agreed to install $460 million in pollution control technologies. This voluntary investment was designed to reduce NOX emissions by 3,263 tons, sulfur oxide emissions by 15,000 tons, volatile organic compounds by 68 tons, hydrogen sulfide by 5.8 tons, carbon monoxide by 173 tons, particulate matter by 226 tons, and other various toxic gases by 75 tons per year. None of these reductions were required by any legal requirement, Carman notes.[26]

In return, environmental groups agreed to a "serious" rather than "severe" designation of the ground-level ozone air pollution problem, Carman states.[27] The agreement had a provision that paid for the purchase of two state-of-the-art air quality monitors so that CIDA volunteers can monitor air quality locally. "Our community has lived under the shadows of these facilities for decades, wondering whether a release is harmless or deadly. [With this equipment] we will now have the means to know, instantaneously, what is in the air and tell the community the appropriate response," Kelley said.[28]

Among the early victories Kelley and his colleagues won was an agreement with Valero/Premcor that the company would reduce its emissions of sulfur dioxide at its Port Arthur plant from 225 tons to 125 tons annually. Flaring is down 20 to 5 percent, and the regulatory agencies are issuing more citations for violations of its flaring rules. BASF and Valero/Premcor have invested in some chemical recovery and pollution control devices. And Kelley also convinced the U.S. EPA to impose a violation notice against the Motiva refinery for failure to report a number of upsets in which toxic chemicals were released.[29]

## New Investment in Pollution Controls

When officials at Motiva Enterprises decided to expand their Port Arthur refinery by 125 percent, making it the largest in the nation, they were required by law to apply for a new air permit for these expanded operations. Their request gave a coalition of activists who wanted to reduce pollution from the plant new leverage.

The team that negotiated with Motiva included Hilton Kelley at CIDA; Neil Carman at the Lone Star Sierra Club; Denny Larson at Global Community Monitor; Jim Blackburn, an environmental lawyer; Alex J. Sagady, a specialist in air permit negotiations; Karla Raettig and Eric Schaffer of the Environmental Integrity Project; Layla Mansuri of Austin; and a coalition of environmental justice activists. Together they blocked the Motiva expansion application for over a year. This regulatory intervention, in addition to a vigorous media campaign, brought Motiva officials to the negotiating table on November 6, 2006, to sign a community enhancement agreement, which would both reduce emissions and provide money for programs to improve safety and offset environmental degradation.

In its application to expand its operations, Motiva promised to install an array of sophisticated air pollution control technologies at an earlier date than previously anticipated, as well as more sophisticated monitoring equipment. Neil Carman from the Sierra Club, giving credit where it is due, notes that Motiva "diligently proposed many positive refinery efforts," including expanding the sulfur recovery operation, installing another flare gas recovery system, no flaring of routine emissions, instituting a carefully controlled shut-down process to reduce emissions, improving VOC leak repair procedures, replacing equipment to lower NOX emissions, covering sludge treatment units, removing old storage tanks from service, improving controls on cyanide emissions, installing storage tanks for VOCs to keep down losses, cogenerating electrical power, and a number of other innovations. With these and other initiatives Motiva "may well become the model refinery in Texas with all these positive efforts," Carman observes in a rare compliment to corporate officials.

While Motiva officials contend that their proposal to install these pollution control devices that go beyond what is required by law came about because they made sense to corporate officials, others are convinced that pressure from environmental groups, federal regulatory actions, and lawsuits had something to do with it. Most of the improvements in pollution controls that were put into place at the Port Arthur Motiva refinery prior to the company's expansion permit request were the result of Motiva's "being busted by the U.S. EPA under a massive, industry-wide investigation that uncovered widespread illegal [refinery] expansions and lack of basic pollution controls that we pushed for," assert Denny Larson of the Refinery Reform Campaign and Global Community Monitor. Motiva's improved pollution controls were proposed only after they had been "hammered on" by activists, he adds. Another motivation for offering the pollution controls was that the Sierra Club–CIDA lawsuit referred to "nonattainment" ozone pollution problems that were bound to surface when Motiva's permit application was examined, he says. Using more temperate language, Neil Carman agrees: "I think Motiva knew we would challenge them, and that may have been a strong reason for proposing the new controls."

Unsurprisingly, Motiva officials see this differently. Rick Strauss, environmental manager at Motiva's Port Arthur refinery, says that the company decided on the pollution control investment well before the environmentalists knew what they had planned. For example, Motiva decided to install a flare gas recovery system, even though it was not required by law, because it had proved to reduce flaring minutes by 90 percent at another facility. "We sold that to the company as the right thing to do," he says. Similar decisions were made to install VOC and NOX pollution controls as a way to limit chemicals that cause smog.

### Community Enhancement Agreement

Hilton Kelley used the leverage with Motiva over their expansion permit application to argue that the company should put money into improving health care and conditions of life for fenceline residents who would be most directly affected by the expansion. "Port Arthur residents on the

West Side are tired of being dumped on and left out of the benefits of these billion dollar projects. If Motiva wants to build the biggest refinery in the nation on top of us then they need to be ready to sign a Good Neighbor Agreement that builds our community and protects our health," Kelley asserted.[30] The expansion of the Motiva refinery in Port Arthur is projected to cause major increases in the amount of pollutants coming out of the plant and drifting into adjacent neighborhoods, he adds.

Motiva spokesman Rick Strauss takes issue with this: "I don't agree with Hilton where he says that our expansion is going to have a significant impact on residents." In fact, for two key pollutants that cause smog, VOCs and NOX the newly expanded facility will actually reduce emissions because old equipment will be replaced with more efficient components, Strauss contends. According to Motiva's permit, the expanded facility will reduce VOC emissions by 7.3 tons a year and NOX emissions by 322.1 tons.

But this accounting does not include the release that will go up, Kelley observes. If the expansion takes place, emissions from the Motiva refinery would increase 31 percent above 2003 levels of 7,340 tons, to 9,632 tons. This would include major increases in sulfur dioxide and particulate pollutants, both of which can cause respiratory harm, Kelley notes. Hydrogen sulfide emissions could also be expected to increase by 1.75 times, he continues. What this means for West Side Port Arthur residents is "that they will have to breathe more dirty air," he warns. Motiva officials concede that the expanded facilities will increase emissions of sulfur dioxide by 2,106 tons, carbon monoxide by 2,048 tons, and particulate matter by 479 tons among other increases.

## No Buyout Option

Kelley argues that in the face of these increased levels of pollution, a good neighbor agreement should include a buyout option so that residents can sell their homes for a fair price; a decrease in emissions, upsets, and flares; an integrated monitoring network; a community environmental education and health center; an evacuation plan; and an independent program to monitor compliance.

Strauss disagrees about a buyout option. He says that Motiva paid for a third-party survey of West Side Port Arthur residents to see what problems they most would like Motiva to work with them to solve. Pollution problems were not high on the list, Strauss notes. Instead, jobs, housing, education, and drugs were problems listed as more pressing than pollution. Residents said they wanted help improving their community, not moving out of it, Strauss continues. Only one or two residents would likely want to move, he adds. This estimate is contradicted by my own interviews with Carver Terrace residents, many of whom said they wanted financial help to move to an area where the air was not so heavily polluted. Confronted with reports that many public housing residents want to be relocated, Strauss notes that the community is divided. "The problem we have as a company is that we cannot do what everyone wants if they all want to do something different," he adds with considerable frustration.

### Two-Million-Dollar Development Fund

Despite these disagreements, Kelley was successful in convincing Motiva to launch a community enhancement agreement that would create a foundation dedicated to the improvement of conditions of life in West Side Port Arthur. As part of that agreement, Motiva officials committed to contribute money to a community development fund on which Kelley will have a seat on the board of directors. Initially Motiva will contribute $2 million to the fund, and, if certain criteria are met, there will be an additional $1.5 million matching grant. The fund will pay for improvements in the quality of housing, as well as fostering new commercial development, social and economic opportunities, and community programs.

"This agreement has the potential of transforming West Port Arthur. It represents the social side of sustainable development, and holds out the hope of environmental and social equity for those living adjacent to the new refinery," Kelley said after the signing. He described the agreement as "unique in the United States." Jim Blackburn, the environmental attorney who represented CIDA, was similarly enthusiastic: "This agreement starts to address a pattern of community neglect and injustice that has existed for decades in West Port Arthur. Our hope is that it will be the first of many steps toward equity for the community." Blackburn lauded the settlement agreement as "the best example of a sustainable development agreement in the United States. . . . The inclusion of a so-cial component makes this settlement agreement both precedent-setting and exciting."

Beyond the good news about Motiva's contributions to a community development fund, there nevertheless remain outstanding health issues not resolved by the agreement. For example, nothing in the agreement commits Motiva to pay for a buyout option for residents who want to move out of the community to avoid the additional pollution that the ex-panded facility will emit. Relocation of residents was brought up during negotiations with Motiva but left out of the final agreement.

Nor does the agreement make clear what will be the fate of the many elderly residents who already use oxygen tanks in order to survive the polluted air in their West Side apartments and the large number of young people who need medication and treatments for pollution-induced and aggravated asthma. Whether these vulnerable populations eventually will be relocated remains to be seen. Although it is currently unwilling to pay to relocate these people, Motiva has offered to help pay for respiratory tests and treatments by committing to donating money to the Gulf Coast Health Center over a five-year period starting with a $25,000 contribu-tion. The money will be used to provide low-cost and no-cost respiratory analysis and treatment for residents who qualify.

While the community enhancement agreement with Motiva is an im-portant victory for Kelley and residents of West Side Port Arthur, CIDA, and the coalition of environmental justice activists who supported the clean air campaign, it remains to be seen how the agreement will be im-plemented and whether the pollution control equipment of the expanded facility will work well enough to improve the air in Carver Terrace apart-ments. However, the agreement does provide Kelley's CIDA with two new handheld pollution monitors, as well as new stationary air monitors. An advanced hydrogen sulfide odor detection device will also be purchased and installed. Funding will be made available for an improved disaster warning system to protect residents in the event of potentially harmful releases of toxic chemicals, as well as funding for better access to health facilities for local residents. Finally, the agreement commits Motiva of-ficials to provide better exchanges of information about the operation of their plant with local residents, including an annual environmental report to the community.[31]

Motiva officials emphasize that their operations are getting cleaner. There has been a significant reduction in emissions from the Port Arthur Motiva refinery since 1990, contends Motiva spokesman Stan Mays: "Millions of dollars have been invested in pollution control equipment at Port Arthur over the past decade. As a result, there has been a 75 percent reduction in emissions since 1990. In 2004 the refinery flared 70 percent less often and 90 percent fewer minutes," and there was an additional 21 percent decrease in 2005. But many of the residents of Carver Terrace say they are still choking on the bad air from surrounding heavy industries. West Side Port Arthur resident Juaniki Conley, who grew up locally, says she suffers from bronchitis, elevated blood pressure, hypertension, and allergies. She also has three children "who have to have breathing treatments." Do these health problems have anything to do with pollution wafting across the fenceline from Motiva and other heavy industries? She thinks they do, and so do a lot of her neighbors.

**Long-Term Struggles**

Without the determined organizing of Hilton Kelley and support he received from many residents, environmental conditions would likely be worse than they are today. Campaigns to clean up the air in fenceline communities such as the one led by Kelley are not won in a day, a month, or even a year. They are often multiyear (sometimes decade-long) struggles that require constant vigilance. Recognizing this, some industrialists and their community relations managers know that if they can just stall resident demands for improved environmental conditions long enough, the grassroots leaders will burn out, and the organization they lead will disappear.

Fortunately some grassroots activists refuse to fade away and refuse to fall silent despite serial disappointments and apparently insurmountable obstacles. In the next chapter, we meet another such environmental justice activist, Suzie Canales, who took up the plight of residents living on the fenceline with Corpus Christi's "Refinery Row," one of the largest concentrations of high-emission petrochemical facilities in the nation. Anyone waiting for Canales to tone her protests down had better not hold their breath.

# 4

## Corpus Christi, Texas: Hillcrest Residents Exposed to Benzene in Neighborhood Next to Refinery

Suzie Canales grew up on Karen Drive in a race-zoned neighborhood in Corpus Christi, Texas, a city whose Latin name means "body of Christ." When her sister died of cancer, Canales became convinced that something was wrong in her neighborhood and began to do some research. After wading through a swamp of city records, Canales discovered that the Cunningham area in which she and her sister were raised was located adjacent to two oil waste dumps later used as municipal garbage landfills. In the 1940s, the area was designated by city officials as "reserved for Mexicans." This follows a documented pattern of placing a disproportionate share of hazardous waste dumps in communities of color.

One of the landfills near the Canales homestead was the forty-seven-acre Greenwood hazardous waste dump. It separated the forty-four homes in the Cunningham residential neighborhood from another race-zoned area designated for "Negroes." Race zoning was subsequently ruled illegal by the courts, but by then, people were already in place, and many of the original low-income and minority residents continued to live in these neighborhoods.

As a child, Canales, whose maiden name is Bazan, was unaware that she lived in a race-zoned area. She attended the Chula Vista Elementary School and Cunningham Junior High School, where she marched in band practice on top of a site she later discovered was a small hazardous waste dump that had been covered over with topsoil and grass. It wasn't until several decades later, after she married, had children, and had traveled with her Navy husband for twenty years to a variety of posts, that she became aware of her family's early exposure to toxic chemicals.

With her husband ready to retire from the Navy and settle in New England, Canales received word from her family in Corpus Christi that her older sister, Diana Bazan, then forty-two years old, was dying with stage-four breast cancer that had metastasized to the brain. Canales returned home to be with her sister, who died on December 29, 1999. She and her sister had attended Cunningham High School, and a number of their former classmates attended the funeral. One after another, they approached Canales to offer their condolences. They also told her that a surprising number of former classmates either had died of cancer or were currently suffering from it. Suspecting that this might be more than a coincidence, Canales began to take notes and keep a list of their names.

The day after the funeral, while cleaning out her sister's home, Canales and other members of her family began to talk about whether her sister's death might have been prevented. "Maybe it was the dumps that made her sick," Canales mused. Her curiosity aroused, Canales placed ads in two local advertising publications, the *Thrifty Nickel* and *Adsack*, requesting that former residents and students from the Cunningham neighborhood call her if they or someone in their family had cancer. After receiving numerous calls, Canales conducted an informal health survey in the neighborhood, which led her to suspect an unreported public health problem.

## Getting Organized

Working with other concerned citizens, Canales and her family founded a grassroots group, Citizens for Environmental Justice (CFEJ), in March 2000. The group began to hold press conferences and appear on television and radio shows to expose the problems being experienced by residents who still lived in the previously race-zoned neighborhoods adjacent to the refineries and hazardous waste sites.

Canales and other members of CFEJ embarked on a lengthy research effort in which they documented the existence of the hazardous waste sites near her childhood home and school, a fact that city officials had previously denied. They also uncovered records of the deliberations of city officials who wrote the race-zoning ordinances that had placed her and her family in one of the middle homes on a street bracketed with hazardous waste sites.

The research Canales and her family did took years and is extraordinarily detailed and comprehensive. Her small apartment is stacked with piles of copied city documents, maps, lengthy scrolls of news clips taped to butcher paper, epidemiological studies, and official records. In addition to searching through the property records at town hall, Canales delved into city planning archives, the Rail Road Commission files, as well as school board meeting minutes looking for clues about where poisons had been buried. What she found was that Nueces County and Corpus Christi were at one time home to an estimated "3,760 wells in 89 oil fields, within a radius of 125 miles of the port."[1] Corpus Christi is pockmarked with hundreds of oil waste pits and dump sites, many of which were later converted into garbage landfills. She also uncovered the fact that oil and gas pipelines ran beneath her old community and school and were routed through the waste dumps themselves. This is a dangerous practice, Canales contends, because materials in the waste dump seep down and corrode the gas pipe, causing leaks and further contamination.

## Elevated Disease Rates

In addition to generating a map of old and new oil industry waste pits, Canales and her family looked for evidence that exposure to toxic chemicals from oil industry waste had caused elevated levels of disease in her community. To this end, she developed a door-to-door informal health survey in the Cunningham neighborhood where she lived as a child. She found a surprising number of cancers and birth defects and a high incidence of hysterectomies among women, some as young as seventeen or eighteen years old. Other health complaints included headaches, nosebleeds, and immune deficiency diseases.

Aware that this kind of informal health study was only a first step, Canales began to look for data about cancer rates in Corpus Christi and to push for new studies. The paper chase proved frustrating. The first studies done by the Texas Department of State Health Services in 2000 showed no high incidence of cancer. But with help from outside experts, Canales did a critique of the study, noting that it was based on only two years of data compared with most state studies that covered five years. "These studies are designed to fail," she observes.

Unwilling to give up, Canales took her complaint to an ombudsman at the federal Agency for Toxic Substances and Disease Registry (ATSDR) operated by the Centers for Disease Control in Atlanta. ATSDR deferred to the Texas Department of Health Services. Convinced that the state cancer data were incomplete, Canales went back to the ombudsman at ATSDR and asked epidemiologists to take a more comprehensive look at the cancer data in Corpus Christi.

A second study, published in October 2001, reported elevated levels of colon, stomach, bladder, kidney and renal, esophageal, breast, and leukemia cancers. State officials said that these were unlikely to be caused by environmental exposure to pollution from refineries and hazardous waste sites, arguing that if the problem were due to environmental exposure, then the increase in cancers should be found across gender and ethnic lines. This reasoning appeared faulty to Canales, who suggested that more men worked in the local petrochemical plants and might have had occupational exposures that increased their cancer rates.

The collection of cancer data by postal code also appeared to Canales to be an inappropriate way to determine whether an environmental cause was at the root of elevated cancer rates in her community. The cancer zip code data are based on where a person died rather than where he or she lived for extended periods prior to death. Canales found fault with this approach, reasoning that obtaining an accurate picture of environmental exposures requires that researchers interview residents to establish where they previously lived. Mounting such a study, however, requires money and the political will to find the cause of community health problems, she points out.

Despite the inadequacies of state cancer statistics, Canales began to find confirmation elsewhere for her suspicions that the health of residents in her county was being impaired by industrial contamination. One source was a study published by Public Citizen, which found that residents of Nueces County had higher death rates from cancers associated with industrial pollution and significantly higher rates of hospital admission for adult and pediatric asthma attacks when compared with the state average.[2]

Further evidence that industrial pollution was a public health threat in Corpus Christi came when the first in a series of studies of birth defects was published starting in August 2001. The most recent study, published on July 7, 2006, revealed that infants born from 1996 to 2002 in Nueces County, where Corpus Christi is located, had an 84 percent higher chance of being born with a birth defect than elsewhere in the state and a 17 percent chance of being born with a severe defect. This caught the attention of the media, and suddenly people began to think that perhaps contamination was causing health problems in the community.

Birth defects are a better indicator of environmental problems than cancer rates because the latency period is just nine months, whereas cancers can have a latency period of ten to forty years, Canales explains. Her research hit home when Canales learned that two of her grandchildren, Justin and Julian, were born with two different heart defects of types found to occur at elevated levels in her community. This made the problem personal and provided Canales with more reason to look for the cause of health problems in her old neighborhood. While health officials were unwilling to make the connection between contamination and community health problems, Canales forged ahead by plotting the location of the waste sites on a city map and then adding known rates of cancer on a plastic overlay. "It matches up," Canales reports: the combined maps demonstrate how closely the location of the waste sites and elevated cancer incidence coincide.

Convinced that her sister had been a "casualty of environmental racism" and suspecting that her grandchildren's birth defects were caused by industrial contamination, Canales broadened her research into environmental pollution in Corpus Christi to include communities located adjacent to the six Citgo, Flint Hills Resources (Koch), and Valero refineries. It was this interest in petrochemical exposures that brought Canales to Hillcrest, located on the fenceline with Refinery Row, where she found the most intense contamination in the city blanketing neighborhoods inhabited by low-income residents and people of color.

## Capturing Evidence

Canales was my guide on a toxic tour of Corpus Christi's Refinery Row, the densest concentration of refineries in the nation. She accompanied me on a meandering drive through a landscape of giant industrial plants including six major refineries owned by Citgo, Valero, and the Flint Hills Resources (Koch) companies, as well as an asphalt plant, a Javelina gas processing unit, and facilities owned by Trigeant and Air Liquide. Two of these refineries—Flint Hills Resources and Valero—earned Corpus

Christi the distinction of being the only city in the nation to have two refineries in the list of top ten emitters of cancer-causing chemicals.

"It's disgraceful what is going on here [in Hillcrest]," Canales says. Flint Hills Resources (Koch) Refinery paid the largest civil and criminal fines in U.S. history, and now Citgo is under criminal indictments. "The injustice goes on and on as people continue to suffer from the refinery and chemical plant pollution and the regulatory agencies ignore pleas for help," she adds.

To make her point, Canales brings me to the home of Horace Smith, who is not shy about lambasting the refinery companies for compromising his health. "The life is being sucked right out of us," says Smith as he sits on a stool in his living room with a clear plastic tube running from an oxygen tank up his bare chest to his nose. "I'm real short of breath, and it [the oxygen] helps me breathe," explains Smith, who lives on Palm Drive in Hillcrest. Just two short blocks from his door and clearly visible from his porch is the huge Flint Hills Resources (formerly Koch) refinery.

"Today it smells like gunpowder," says sixty-one-year-old Smith, whose wife lies in bed at midday too debilitated by the fumes to get up. "Last night was bad," he continues. His wife was up all night coughing and had to borrow his oxygen breathing apparatus, he explains. "I don't mind telling you I cried last night because there was nothing I could do for my wife."

"We are stuck here. This is all we have," Smith adds, gesturing vaguely at the small wooden house where the smell of benzene and sulfur dioxide permeates the room, stinging the eyes and irritating the back of the throat. The window screens are lined with tinfoil where the glass is missing and four oxygen canisters lean against the wall. Outside is a neatly kept flower garden, and guard dogs are penned up in the back. "We live on government money, and that's not much," says Smith, who wants the Citgo and Flint Hills Resources refineries to buy him out at a price that would permit him to own a house on the other side of a nearby freeway. "That would be far enough [from the fumes]," he calculates.

On the two streets between his property and the refinery, most of the homes have been torn down, and just a few holdouts remain—modest buildings surrounded by a vast lawn. The refineries buy up the properties at fire sale prices ($30,000 to $50,000) from people desperate to move. The prices are low because no one else wants to buy into a community where the air is heavily contaminated and where, as Smith points out, the flares are so bright at night "that you can read a book inside without turning on the light or pick up a nickel off the street."

For a number of years, Smith, a retired truck driver who hauled loads in and out of the plants near his home for years, attended meetings with refinery officials, but they never delivered on the promises they made, he charges. "You get ear rot listening to that. I can't hack it. It don't pay to crank up the car to go listen to that."

Asked if his doctor is willing to say that he should be relocated away from the source of the fumes for the sake of his health, Smith laughs. "The doctors are paid under the table, and the city council is corrupt. You don't have a chance here." Now city officials are planning to locate a sewage treatment plant a few blocks away. Officials don't want the facility too close to the baseball stadium "where people with money might smell it," he claims. Instead they are going to build it in Hillcrest, a low-income, African American community that is already burdened with heavy emissions. "What they are saying is that we don't matter," says Smith with a combination of sadness and outrage in his voice. Horace Smith died in December 2008.

## Toxic Chemical Releases

Horace Smith's conviction that he was being poisoned by the refinery seen from his front door is given substance by publicly available data about emissions from the two Flint Hills Resources (FHR) refineries. First bought in 1952, the Koch Petroleum Group LP built the west side refinery in 1981 and a second east side plant in 1995. In 2000 a criminal indictment was brought against the company for hiding and misrepresenting benzene pollution, and a year later, corporate officials paid a $10 million fine and agreed to invest another $10 million in pollution control projects.[3] The Koch refineries were renamed Flint Hills Resources in 2002, the year in which its two refineries released 8,729,324 pounds of criteria air contaminants and an additional 584,539 pounds of toxic chemicals.[4]

Some of the accidental releases from these FHR refineries are clearly dangerous to the health of nearby residents. For example, on August 28, 2003, FHR West released 130 pounds of benzene and 4,203 pounds of hexane. "Benzene is a recognized carcinogen that causes leukemia, a developmental toxicant, and a reproductive toxicant. It is also a suspected toxic to the cardiovascular and blood system, endocrine system, gastrointestinal system or liver, immune system and skin or sense organs."[5] Breathing air laced with hexane is not much better for humans and might help explain a number of the symptoms from which Smith and his wife suffer. "Hexane is a suspected developmental toxicant, neurotoxicant, reproductive toxicant, and respiratory toxicant. The symptoms of hexane exposure include irritation of the eyes and nose, nausea, headache, numb extremities, muscle weakness, dermatitis, dizziness, and chemical pneumonitis."[6]

## Health Impact

A former Hillcrest resident who thinks her health has been impaired by exposure to contamination is forty-four-year-old Gwendolyn Nickerson, who now lives in public housing, next to a malodorous sewage treatment plant, in a high-crime area known as "The Cut." Prior to moving to her current residence, Nickerson lived with her mother and daughter in Hillcrest on the fenceline with the Flint Hills Resources and Citgo refineries. It was during this period, in 1999, that she was diagnosed with scleroderma, a rare autoimmune system disease that causes her body's immune system to attack her tissues creating a scarring, or fibrosis, of the skin and organs.

"Too many people are sick of rare diseases that no one ever had before," says Nickerson, an emaciated woman who says she has lost a lot of weight since being diagnosed. One of her hands is curled inward from the tightening and hardening of the skin, making it impossible for her to grasp objects or keep her jobs at the Family Dollar and Whataburger. She also suffers from acute pains to her extremities whenever it is cold, a result of Reynaud's phenomenon, in which spasms of the tiny blood vessels interrupt the supply blood to the fingers and toes.

"They say it can't be cured, just treated," says Nickerson, fighting back tears as she sits in her three-story walk-up apartment where she lives with her eighteen-year-old daughter who has helped her deal with her disease since she was a ten-year-old child. Nickerson lost all of her teeth as a result of scleroderma, has difficulty walking any distance, describes herself as "short-winded," and sometimes can't get out of bed or feed herself.

The causes of scleroderma are unknown, but two current theories suggest the illness is either passed from one generation to the next genetically or is caused by environmental factors. Since no one else in Nickerson's family has had similar problems and she has been exposed to heavy concentrations of petrochemical toxins, the environmental exposure could have caused or triggered the disease. Specifically, benzene and butadiene, both chemicals put out by nearby refineries, are listed in the medical literature as having "good" evidence that exposure to these chemicals is linked to the disease. "It makes you wonder," Nickerson says about whether her disease might be related to exposure to chemicals from the refineries.

### "Ma, I Can't Breathe"

A number of other Hillcrest residents are also pretty sure that their respiratory problems are related to fumes coming from the refineries. Since moving to the home of her father in Hillcrest two blocks from the refinery, Elizabeth Sobrano, thirty-one years old, says that her eight children have been chronically sick, and a number of them have asthma and respiratory problems. She has taken her children to the doctor with breathing problems so many times that the doctors avoid seeing them. Nevertheless, Sobrano keeps going back: "I can't help it. When my eighteen year old says: 'Ma, I can't breathe,' I have to take him."

When her doctor suggested to one of her sons that his problem breathing was caused by his mother's failure to keep their house clean, her son objected and said that his mother did a good job of housekeeping. The doctor's comment made Sobrano mad: "I'm poor. I don't have everything that you find in a rich person's house, but I keep what we have clean." As an alternative explanation, Sobrano suggested to the doctor that the fact that her father had to use an oxygen tank to help him breathe and that her children all had respiratory problems might have something to do with living next door to a petrochemical plant. "You don't live inside the refinery," the doctor replied, according to Sobrano, who felt his response failed to take into account the fact that fumes from the plant escape across the fenceline into the residential neighborhood.

Since moving to Hillcrest, Sobrano has developed a cough and some-times borrows one of her children's inhalers to help her breathe. She is also often depressed and without energy. "And that is not me," she con-tinues, explaining that she is normally more upbeat and energetic. "We never used to use medications before we came here," she adds. Closing the windows and turning on the air conditioning helps her three children with asthma breathe, but she can't do that all the time because running the air conditioner is expensive.

"The odors come from the plant, and you can smell it in the house. It gets into the walls and into your clothes. I had some clothes I had to throw out, it was so bad. And when my mother-in-law came to the house, she said it smelled ugly. Now no one wants to visit," she says. "I'm wor-ried my dad will die," says Sobrano, whose mother died in her sleep recently. "If Dad died, I wouldn't know what to do because I am a single parent. Sometimes I go outside and cry because I don't know what to do. It is too expensive to move," she explains. (Sobrano's father died a year after this interview.)

Not far from where Sobrano lives is the home of Janie M. Mumphord, seventy-five years old, a woman of erect bearing who sits inside her neatly kept home filled with a chemical smell akin to an insecticide. The odor stings the eyes and makes one light-headed. Mumphord's husband lived in this house for years on oxygen and finally succumbed to emphysema. "I didn't have asthma until after I moved here in 1969," says Mumphord, who also suffers from high blood pressure. During one of the explosions at the nearby refinery, she was so frightened that she became ill and was taken to a hospital. "I looked out of the back of the house and saw black smoke and fire leaping high into the air. Now I sleep in fear that it will happen again. I pray every night for God to keep back the fires," she says. "The doctor told me to move out for my health, but we are retired, and we are stuck here surrounded on three sides by refineries."

Some neighbors have sold their houses for $45,000, but Mumphord refuses to "give my house away" for a sum that would not permit her to buy a home in an unpolluted neighborhood. "A whole lot of people here are dead from cancer or respiratory disease," she continues. When refinery officials say that they know what it is like living here, she is quick to disagree. "No you don't," she tells them. "You stay in my house, and

I will stay in yours, and then you will know." "These fumes are killing us, but I don't think they care," she concludes.

## Sermons on the Evils of Pollution

In the early 1990s, Reverend Roy Malveaux, a Baptist minister, organized People Against Contaminated Environments and preached sermons denouncing the harm done to the community by the refineries. Malveaux left town after industry officials offered jobs to the sons of the deacons in his church and subsequently threatened to fire them if Malveaux was not sent packing, claims Denny Larson at Global Community Monitor. Other activists were silenced with payoffs, Larson contends.

Reverend Harold T. Branch also decried the hurtful impact of refinery contamination on members of his Hillcrest congregation. Once, Branch received a call from officials at the then Koch (now Flint Hills Resources) Refinery asking him to talk with them about his sermons on the contamination. When he showed up with half a dozen other members of the community, however, suddenly the corporate officials were no longer interested in a discussion and sent them away, he remembers. That was the last real chance for a dialogue, he says.

Branch, now eighty-eight years old, moved to Hillcrest in 1955 from a home in The Cut. "Moving to Hillcrest seemed like an upgrade to us," he recalls. Only later, as the refineries began to expand and engulf the north side neighborhood he had moved into, did Branch and other residents understand why whites had moved out. "A lot of my congregation died of cancer. Most of the men have died—only two are left," Branch observes. "My wife suffers from dizziness, and benzene causes it," he continues. To prove this, Branch reaches into his wallet and carefully unfolds an old, fragile newspaper clipping that lists dizziness as one of the impacts of benzene exposure. "They are going to wait until we all die before they do anything," Branch predicts.

A former member of the city council from 1971 to 1974, Branch was involved in negotiating contributions to the city coffers from the refineries. Twenty years later, he and members of his congregation joined a class action lawsuit against the refineries asking for damages and funds for relocation. Residents of two fenceline communities, Oak Park Triangle and

Hillcrest, joined the lawsuit over the numerous explosions, groundwater contamination, and high levels of benzene and other pollutants emitted by the refineries. Citgo officials were reported to have set aside $17 million to pay for relocating affected residents, but an inexplicable judicial decision provided for the relocation of only the residents of Oak Park Triangle, not those of Hillcrest. What became of the $5 million set aside for Hillcrest residents remains a mystery today.

**Elevated Exposure**

The Dona Park neighborhood, where Zelma Champion and Teddy Murch have lived for some thirty years, has a long and well-documented history of having been on the receiving end of heavy contamination from its industrial neighbors in Refinery Row. When Champion moved into the area, some light industries were nearby, but over the years, these industries expanded and began to handle more toxic chemicals and heavy metals she explains. For example, from 1942 to 1985, ASARCO, a mining, smelting, and refining company, operated a zinc refinery before one of its subsidiaries, Encycle, turned it into a hazardous waste management facility. The Texas Natural Resource Conservation Commission, later renamed the Texas Commission on Environmental Quality (TCEQ), documented numerous environmental health violations at this facility, including the illegal storage of military chemical warfare agents, as well as the deposition of "blue rocks," a highly toxic material extracted from their hazardous waste processing filters. Some of these contained arsenic concentrations as high as 35,870 ppm.

Between 1994 and 1997, Encycle illegally discharged arsenic, lead, selenium, and mercury into their wastewater eighty-eight times, agency records show.[7] In 1994 TCEQ employees sampled the soil in the yards of residents in Dona Park and found dangerously high levels of cadmium, lead, and zinc. Encycle cleaned up fifteen properties but left others untouched. Testing was done again in 1998, but the results were kept secret. A long struggle ensued to force state officials to release the data. After legal pressure was applied, it was revealed that twelve of nineteen additional properties were contaminated with lead and cadmium. ASARCO lawyers argued that the lead contamination came not from their emissions but rather from lead paint used on the housing.

Canales does not buy the argument that sources of pollution other than the refineries and chemical plants are the problem. Most of the residents of Hillcrest are people of color and low-income residents, she notes. Many of them moved to the area many years ago when race-based zoning was in effect and either have stayed or passed their homes to their children. These people are trapped next to these heavily polluting industries because they cannot afford to move elsewhere, she observes. "It's not safe to live there [in Hillcrest]," she continues. A collaborative effort by the city, industry and the community is needed "to right a wrong, end the environmental racism that was set in motion by racial zoning of the past," she says.[8]

**Inadequate Regulatory Response**

More recently, state environmental regulators set up a number of monitoring stations along Refinery Row to detect toxic chemicals coming across the fenceline, but they were not using readings from this equipment to trigger enforcement actions or notify residents of dangerous emission levels, Canales says. Instead, they claimed "there were no health concerns when their monitors were indicating otherwise."

The TCEQ maintained a monitor on John's Street in Hillcrest that found very high levels of benzene, but state regulators never took any enforcement actions based on monitoring, Canales claims. Furthermore, a lot of monitoring is based on averaging, and residents do not breathe in averages, notes Neil Carman, a former state regulator who now heads the Lone Star Chapter of the Sierra Club's air quality program. "If traffic cops gave speeding tickets based on averaging, no one would get a ticket because low speeds would cancel out the high speeds," he observes.

One particularly egregious example of state regulators' failing to use their monitors to protect the community began in 1997 when an air pollution monitor was installed at the intersection of Buddy Lawrence and Huisache streets in the midst of Refinery Row near the Oak Park Triangle neighborhood. From the time it started collecting data, it registered high levels of benzene, a known carcinogen, says Canales. In fact, between 2002 and 2005, benzene readings were either the first, second, or third worst in the state. These readings were annual averages and do not take into account the high spikes of benzene releases. Canales is particularly outraged that nothing was done with these data because there was a school bus stop at this same intersection where children waited for a ride to Gibson Elementary School. "No one has ever looked to see what happened to the health of those children who used to wait for the bus there," Canales adds.

In an analysis of the monitoring data collected at the Huisache intersection from 1997 to 2003, Wilma Subra, an independent chemist who helps grassroots groups understand complex regulatory and health issues, notes that for every year during that period, benzene levels exceeded the Texas Annual Health Effects Screening Level of 1 ppb, spiking at 3.8 ppb in 1997 and rising above 2 ppb from 2001 through 2003. The state's twenty-four-hour screening level of 4 ppb was also exceeded at this intersection on specific days from 1997 to 2003, rising to 27.65 ppb in 1997 and 28.54 ppb in 2002. These readings are seven times the acceptable standard, Subra continues and concludes, "The concentrations of benzene in the ambient air over such a long time period is unacceptable."

Canales tried to convince TCEQ regional director Buddy Stanley to come to the community and discuss the high levels of benzene detected at the Huisache intersection, but in a letter to her on March 19, 2004, he declined, arguing that "none of the monitors have detected benzene levels of a health concern."

**Capturing Evidence**

Fed up with what she saw as the do-nothing approach of state regulators, Canales started her own monitoring in August 2003. Denny Larson, director of Refinery Reform Campaign and Global Community Monitor, came to Corpus Christi and trained Canales and a few other residents in the use of citizen air monitoring devices made out of plastic buckets, air pumps, and special bags in which an air sample can be trapped and then sent to a lab for analysis. "We documented gases outside the refinery fenceline" using the new equipment, Canales says.

A year and a half later, on February 11, 2005, Hilton Kelley, the fenceline activist from Port Arthur, Texas, arrived in Corpus Christi with a $35,000 Cerex ultraviolet "hound" monitor that gives a computer readout of toxins in the air. "Before that when Cindy [Canales's sister and cofounder of CFEJ] and I would patrol, we'd sniff and wonder what the heck we were smelling that was so bad. Then when Hilton brought the Cerex, we actually saw the names of the gases come up on the laptop screen. We were in awe of the technology. Finally, we were seeing identified what we were smelling," Canales recalls.

One of the areas she monitored was the Buena Vista Mobile Home Park, where sixty trailers are parked next to the Valero East refinery tank farms and Javelina gas processing plant. It was there that Canales, using a Cerex monitor, captured a reading of 42 ppb of 1,3-butadiene, a recognized carcinogen that also causes respiratory, nervous system, and liver problems. Later, on November 14, 2005, Canales recorded readings of benzene releases from the same refinery spiking at 114.09 ppb at 3:38 p.m. and at 42 ppb at 7:56 p.m.

"We found what we already knew: that refinery emissions were crossing the fenceline into the community contrary to what industry and the TCEQ were saying. The state, federal, and corporate sources failed to uncover it because they didn't want to. They have the money and capability to uncover it, but they are not into protecting human health; they are into protecting industry," Canales asserts.

A recent study also suggests that the cumulative exposures of Hillcrest residents to industrial toxins have left them with elevated levels of benzene in their bodies. The study—conducted by the Texas A&M School of Rural Public Health, the Coastal Bend Health Education Center, and Citizens for Environmental Justice—found that blood and urine samples of Hillcrest residents contained 14 times more benzene in their bodies than did a sample of gas station attendant workers studied in Mexico and 280 times as much benzene in their blood and urine compared with people in the general population in the United States.[9]

"The results are in; the results are alarming and disturbing. We hope this will prompt our city leaders to do the right thing: instead of talking of revitalizing the community, bring more children into an area that is highly toxic, they should be working to relocate the community in a manner that is fair to the people," Canales observes. The refineries purchased a number of residences in the 1990s but terminated the program in 1999. Canales thinks the program should be revived.

Bird Deaths Lead to Overdue Regulatory Action

When migratory birds were found dead in open oil tanks, regulators were forced to grapple with the problem posed by benzene releases coming from Refinery Row. "Why did it take the birds dying to make the agencies force Citgo to comply with regulations?" Canales asks.[10] This seems a reasonable question given that Corpus Christi held the dubious distinction of having either the first or second highest emissions of benzene in the nation from 2001 to 2004.[11]

One massive source of benzene releases came from two uncovered Citgo oil-water separator tanks, which were found to be holding 4.5 million gallons of oil during an unannounced inspection in 2002. These tanks should have had fixed or floating covers, but they did not. On August 9, 2006, a federal grand jury issued a ten-count indictment against Citgo for violating the National Emission Standard for Benzene Waste Operations for releasing tons of uncontrolled benzene during the nine-year period January 1994 to May 2003 in violation of the Clean Air Act and the Migratory Bird Treaty. Citgo officials ultimately paid two fines: the first for $725,000 in 2002 (later reduced by half in return for voluntary contributions to conservation and pollution prevention efforts) and the second for $1.74 million in 2004.

Why did it take so many years for state regulators to require Citgo to stem massive releases of benzene? Neil Carman blames it on "a coziness" in the relationship between regulators and the industry they are supposed to be regulating. This "coziness" has been endemic to the state's environmental regulatory system since the state adopted a "cooperative enforcement policy" under the administration of President George W. Bush, observes Victor Flatt, who holds the Chair in Environmental Law at the University of Houston Law Center.[12]

But the problem for residents facing contamination from facilities along Refinery Row goes well beyond a couple of uncovered oil tanks. For example, one of the six refineries, Valero Energy Corporation, has contributed heavily to air contamination in Corpus Christi over two decades. In 2003 the Valero West refinery emitted 7.5 million pounds of criteria air pollutants, including nitrogen oxides, sulfur dioxide, carbon monoxide, ozone, particulate matter, and lead. The refinery released an additional 268,904 pounds of toxic chemicals into the air—a category even more harmful to health than the criteria air pollutants. This record placed Valero West in the top 10 percent of the "dirtiest facilities" in the United States and in the top 20 percent of the worst emitters of recognized cancer-causing and birth defect-causing chemicals[13]

While it is notoriously difficult to prove in court that these releases are the cause of health problems in surrounding communities, common sense suggests that they are. Also suggestive are data showing that following heavy accidental emission events, attendance drops at nearby schools. For instance, two days after the facility released 54,523 pounds of sulfur dioxide into the air on May 20, 2004, attendance dropped at Gibson Elementary School, located within a two-mile radius of the plant.

### Help from Outside

Canales's organizing activities eventually made newspaper headlines and attracted the attention of organizers in the state and national environmental justice movement, who have begun to provide her with help. Denny Larson from Global Community Monitor, who helped her set up the bucket air monitoring, provided funding for Canales to continue her organizing work, paid for the analysis of air samples she collected, and bought her a second-hand 1994 Jeep Cherokee when her old car fell apart and she was unable to get to Hillcrest and patrol the perimeter of Refinery Row. Without the financial assistance and advice from Larson, Canales says, she would not have been able to continue as an organizer.

Canales also received help with some of the hardware needed to run an effective campaign. Anne Rolfes, founder and director of the Louisiana Bucket Brigade, who has organized fenceline efforts in Norco and Chalmette, Louisiana, near refineries in those communities, donated a computer, as did Jennifer Carraway, who worked at Public Citizen at the time and has since moved to Environmental Defense. More recently Stan Johnson at the Environmental Support Center provided a new laptop computer, a digital camcorder, a digital camera, a digital projection screen, and a printer/scanner/fax machine. Canales put this equipment to good use on April 1, 2006, when she filmed heavy particulate pollution coming off the coker unit of the Citgo West facility. The film shows the lens of the camcorder gradually becoming occluded with soot. She subsequently showed this footage to TCEQ regulators at a regulatory hearing on June 8, 2006.

## Suspicious Activity

While engaged in filming the pollution emitted by the Citgo facility, Canales was reported to the National Response Center for "suspicious activity" and was subsequently interviewed by agents from the Federal Bureau of Investigation in her apartment. What was she doing filming a petrochemical plant? agents asked. Canales told them she had been patrolling Refinery Row for years monitoring environmental problems associated with the operation of the plants and produced voluminous documentation to prove it. She found it strange that she had been reported for suspicious activities after having patrolled along the roads around the refineries for years and wondered why she hadn't been reported before. Whatever the reason, it was unnerving to have the FBI come to her home, Canales says, but she remains unwilling to stop patrolling Refinery Row to document evidence of pollution.

Moving beyond simply capturing air monitoring data, Canales began to intervene in the air permit process of a number of refineries. When the Citgo East plant next to Hillcrest proposed an expansion of its fluidized catalytic cracking unit, which would have increased production by thirty-five hundred barrels of oil a day and produced an additional five hundred tons a year of sulfur dioxide emissions, Canales contested the expansion at a permit hearing in 2005 and two years later was informed that the TCEQ had turned down Citgo's permit application because of the company's unwillingness to install the best available pollution control equipment. Since contesting an air permit is an extremely technical endeavor, Canales sought help with it from Neil Carman at the Texas Lone Star Chapter of the Sierra Club, Eric Schaeffer at the Environmental Integrity Project in Washington, Denny Larson with Global Community Monitor, and Enrique Valdivia at the Texas Rio Grande Legal Aid Society. "It was a huge environmental justice victory that made all the newspapers," reports Canales, who was honored in Washington, D.C., with the Congressional Hispanic Caucus Institute Award for Outstanding Achievements in Environmental Justice.

## Fines Don't Go to Help Affected Communities

Canales also protested the fact that on the rare occasions when regulatory agencies imposed fines on the refineries for accidental emissions, the money collected from these fines seldom ends up in the affected community. In the tortured jargon of the regulatory bureaucracy, there is something known as supplementary environmental projects (SEPs), which corporations caught polluting can voluntarily agree to fund in return for reduced cash fines. Ideally these projects should provide direct benefits to residents in the communities where the emissions have the most powerful impact. However, Canales found that SEPs funded land conservation acquisitions, bird nesting sites, fire department equipment, conferences in Houston, programs to teach water conservation, and a vehicle emission sensor program. "SEP's are routinely awarded for wildlife and other projects; left out of the equation is the impacted community itself that had to bear the burden to their health from these violations," Canales writes.[14] As an alternative, Canales suggested that an SEP with Citgo be amended to provide funding for one of four proposals put forth by Hillcrest residents: a body burden study of pollutants found in residents, a respiratory health survey, compensation for property depreciation, and improvements in the emergency planning and preparedness along the fenceline. Citgo refused to fund any of these.

All of Canales's activities criticizing the amount of pollution emitted by Refinery Row industries and holding press conferences revealing regulatory inaction have not made her popular with Port of Corpus Christi industrial managers. In a telephone conversation, one of the officials told Canales: "You have stirred them up. . . . You've set yourself up as their archenemy" and "you wouldn't be anyone's first choice for a prom date among that crowd." It isn't fun knowing that the owners of multibillion-dollar industrial operations don't like you, says Canales, who thinks that her phone is tapped and that she is sometimes followed.

But her anxieties over the possible consequences of being a thorn in the side of powerful petrochemical executives has not deterred her from her work delving into the details of how her community is being poisoned by nearby industries. With remarkable perseverance, Canales has made herself expert in the minutia of how to challenge refinery air permits, how to operate sophisticated air pollution monitoring devices, and how to read a regulatory report. In other words, she has learned to do the job that state and federal regulators should have done.

Looking around at the piles of research reports and air pollution monitoring data stacked on the floor of her apartment, the question is why Canales had to do this work largely on her on her own, without compensation or help from government agencies. Why is the uncovering of illegal levels of exposure to toxic chemicals being left up to this grandmother of two ailing grandsons who has lost her sister? Why does the government not have its own undercover environmental inspectors doing this vital work?

Since many government officials are either not interested in catching polluters who emit dangerous levels of toxins into residential areas or lack the resources to carry out these types of investigations, it is left to fenceline groups to figure out how to do it. But not all grassroots antipollution campaigns evolve in a similar fashion. In Ocala, Pensacola, Port Arthur, and Corpus Christi, local residents perceived an environmental health threat, organized themselves, and then reached out for help from other environmental justice activists and experts. In some communities, however, as we will see in the next chapter, it takes organizers from outside to come into the community and perform environmental audits and door-to-door environmental awareness campaigns before local citizens are emboldened to join together and protest a local source of pollution. This is what happened in Addyston, Ohio, where Ruth Breech and a cadre of organizers built on local concerns about emissions from a plastics plant and eventually won an increased investment in pollution control technology.

# 5

## Addyston, Ohio: The Plastics Plant Next Door

Bernard "Buzz" Bowman Jr. is proud of his antiques collection. He has three thousand antique toy cars and trucks on shelves in his basement in Addyston, Ohio. And this is just the tip of the iceberg. His yard is full of dozens of old-fashioned gasoline pumps, and displayed in his basement is a lovingly restored one-horse sleigh, upholstered in crushed red velvet and sporting a black convertible canopy that snaps into place with well-oiled precision.

"We're on the map," says seventy-three-year-old Bowman, pointing out a designation of his museum on a local tourist handout. He also has a write-up in a book that features attractions along Route 50, the winding River Road that snakes through this West Side factory town located on the banks of the Ohio River twelve miles from downtown Cincinnati on the Ohio-Kentucky state line.

Addyston, population twelve hundred, is also on the map because of toxics problems that state officials and residents say come from the 130-acre Lanxess Corp. plastics plant (formerly owned by Monsanto and Bayer) built across the street from the elementary school and just a few blocks from the Bowman residence. In December 2005, the school was closed after elevated levels of butadiene and acrylonitrile were discovered by monitoring equipment installed on the roof of the building.

"This is cancer valley," asserts Buzz Bowman, whose wife, Carol Bowman, died recently of lymphoma. Her oncologist described her lymphoma as "an environmental cancer" but refused to speculate as to whether it might have been caused by living next to a plastics plant all her life. His reply infuriated Buzz's daughter, Lynn Bowman, fifty-two years old: "Do they think we are stupid? My mother died of an environmental cancer, and we live right across from a chemical plant. In years to come, they will find out that it was caused by the plant," she predicts. "The air pollution had a lot to do with it. It killed my mother," she charges. "It is amazing what big corporations get away with." Before she died, Carol Bowman, known locally as the "candy lady" for the fudge she baked, joined other residents in protesting chemical releases from the plant. She told her daughter that the protests came too late to prevent her illness, but perhaps they would save others.

## Large Volumes of Toxic Chemical Releases

Sitting on the couch in her living room, Lynn Bowman describes what it is like to live next to a plant that put out 224,000 pounds of toxic chemicals, including butadiene, styrene, and acrylonitrile in 2005, and legally released 1.6 million pounds of particulates and toxic chemicals, including 813,000 pounds of sulfur dioxide, 370,000 pounds of nitrogen oxide, 102,000 pounds of volatile organic compounds (VOCs), and 233,000 pounds of particulates in 2003.[1]

"The smell bothers me. I get allergies, and I can't breathe through my nose. On bad days, it just about chokes you. It burns my eyes and nose, and I get frequent nosebleeds. But when we go for a trip and get out of town, it clears up until we come back," Lynn Bowman notes. Releases from the plant tend to be most intense at night when she gets off work late from a job she holds in town, Bowman reports.

"They think we are a bunch of ignorant hillbillies," Bowman says with anger rising in her voice. "Well, don't let this blond hair fool you. I'm no fool. I know what is going on here. The smell from the plant makes us sick. It gives me a headache every day of my life," she adds. Bowman says she has to keep the windows of the house closed because of the air pollution from the plant: "It keeps us from going out sometimes." And periodically a white dust or a kind of plastic sap that is hard to scrape off covers her car.

Ironically, when an inspector from the U.S. Environmental Protection Agency visited the Bowman home, it was not to discuss air pollution coming from the Lanxess plant. Instead the regulator wanted to know if Buzz Bowman's antique gas pumps were connected to an underground gas storage tanks. After Bowman demonstrated that the gas pumps were antiques and connected to nothing, he suggested that the regulator could

more profitably spend his time looking into the source of the odors coming from the Lanxess plant across the road.

The Bowmans are not alone in worrying about health effects from
the plant. "This is a lovely community that is being poisoned," observes Betsy Eckert, sixty-four years old, whose family has lived since
the 1820s a few miles from the plant in Sayler Park.[2] Eckert, the Bowmans, and a number of other residents joined together to form the
West Side Action Group to force Lanxess to appoint a new plant manager, invest in pollution control technology, and reduce emissions from
the plant.

### Getting Organized

Ruth Breech, an organizer for Ohio Citizen Action (OCA), says she
wants to make public the "dirty little secrets" about how industrial polluters cause health problems in adjacent fenceline communities such as
Addyston. A tenacious, high-energy community activist, Breech says she
wants to "tell the untold story" of the people who suffer in silence in
these communities where it is not common practice to speak out about
the odors from the plant. Addyston is a largely white, working-class company town where 97 percent of the tax revenues come from the Lanxess
plant and most residents have family members or friends who work at
the plant, she continues. Criticizing Lanxess for its emissions is seen as
threatening local jobs. "It is a culture where it is not accepted to rock the
boat," Breech observes.

OCA, an environmental watchdog organization with 100,000 dues-
paying members, targeted Addyston and the Lanxess plant for a citizen
antipollution and good neighbor campaign because of the unusually high
incidence of asthma and cancer in the community right across the street
from a plastics plant that was emitting large quantities of cancer-causing
chemicals, Breech explains. Lanxess was also the source of more accidental releases than other companies its size and was one of the top five
plants in the county in terms of the toxicity of the chemicals it used and
the proximity of residents to the plant, she says. "I wanted to see the
community get its act together and demand that the company not dump
a large volume of chemicals, only to tell residents five days later that they
had been gassed," Breech continues.

By chance, Breech knew the Bowmans; she had grown up in Cincinnati where her father had been the place kicker for the Cincinnati Bengals. In September 2004, Breech began to visit Addyston two to three times a week with a cadre of canvassers ringing doorbells and asking residents what it was like living in such close proximity to the plastics plant. The canvassers listened to stories about the persistent odors from the plant, the dust that collected on their cars, and the large number of illnesses in the neighborhood. For most residents, it was the first time anyone had asked them whether the odors from the plant were bothering them, and it got them talking among themselves. Collecting the stories permitted Breech to put a human face on the impact of the plant's pollution on local residents, she observes.

### Rude Awakening

Three accidental releases from the Lanxess plant in late 2004 and early 2005 spurred Addyston residents to organize a protest against the odors and chemicals coming from the plant. The first of these occurred October 2–4, 2004, during the town's Octoberfest, when residents held an annual barbecue at the school playground across from the plant. The emissions included twelve hundred pounds of acrylonitrile, a substance classified by the U.S. Environmental Protection Agency as a probable cancer-causing agent. Laboratory experiments show that rats exposed to the chemical have an increased incidence of tumors. Short-term human effects include headaches, nausea, nervous irritability, and kidney irritation. Poisoning that occurs through longer-term exposure causes "limb weakness, labored and irregular breathing, dizziness, impaired judgment, cyanosis, nausea, collapse and convulsions."[3]

In addition to the release of acrylonitrile, 34 pounds of butadiene and 387 pounds of styrene were released.[4] Company officials explain that the emissions escaped from tiny cracks in a duct that normally conducts the gas to an incinerator within the plant. Lanxess officials reported the release to the Hamilton County Department of Environmental Services, as required by law; they did not, however, inform residents about it. Not until late November did Breech find a reference to the accident in public records and alert town officials. A previous leak of 371 pounds of acrylonitrile over a five-minute period in June

1999 had sickened two employees and resulted in a civil county fine of $37,500.[5]

Jean Owens, an Addyston resident who lives across Route 50 from the plant on Sekitan Street, describes acrylonitrile as having a sweet smell. "It just turns me inside out," she told a reporter. "It's like I can't breathe and I can't move. I do move but I'm just moving like someone who has something really wrong, like cerebral palsy, or something like that. . . . It's just a horrible feeling."[6]

Ken Perica, the plant director of health and safety, assured local citizens that their safety was not endangered by the release. He said that the chemical release, which occurred over a forty-nine-hour period, would have been very dilute by the time it escaped from the plant. He also noted that ten employees working near the leak showed no ill effects. But Perica's assurances were overshadowed by a second accidental release that occurred on December 15, 2004, when Addyston residents were told that they had been exposed to seven hundred pounds of acrylonitrile as a result of a worker error. Once again, the plant manager at the time, Bill Ward, said that air samples showed no "problem with the gas leaving our plant." However, Peter Sturtevant, an enforcement officer with the county, suggested that the gas might have carried over the company fenceline without tripping the air monitoring sensors. This time Lanxess officials promptly notified the town mayor and other public officials of the release.[7]

A third release two months later, on February 23, 2005, received major coverage in the local press and suggested a problematic pattern of emissions from the plant. This time the 750 pounds of butadiene had been unintentionally released. The company is required to report releases of 10 or more pounds of the substance, which is known as an extremely toxic, cancer-causing gas. "Obviously, that's not acceptable," observed Jay Richey, Lanxess vice president and general manager for North America. Richey promised that policies and practices at the plant would be reviewed promptly. Company officials noted that none of their workers had been made ill by the release and suggested that the wind was blowing away from Addyston at the time of the release.[8] "These unintended releases did not expose our employees or anyone in the community to harmful levels of chemicals," he asserted.[9] An Ohio Department of Health study, which used computer modeling and company data, agreed that the three releases did not pose a significant health risk to nearby neighbors. Only the December 15 release "rose to the level of potentially causing headaches, sore throat, or watery eyes," the report suggested.[10]

Ruth Breech at OCA argued that the study was suspect because it relied on company data. "These accidents are so consistent that everyone needs to be more proactive" to prevent future releases, she commented.[11] "The Lanxess Plastics Plant in Addyston has chronic, serious problems, including 107 accidents in 2004," Breech wrote in a guest column for a local newspaper.[12] She views the series of accidents at Lanxess as dangerous to the local residents. "I definitely believe there is wrongdoing that is going on here and it needs to be changed," she told a TV reporter.[13] "If the company can't operate responsibly then they should not be permitted to operate at all," she asserts. But Breech is not trying to close the plant: "We want them to stay here. We don't want them to shut down. That is not our intention. We are here to clean them up. We want them to be good neighbors for a very long time."[14]

Mike Kramer, environmental enforcement supervisor at the Hamilton County Department of Environmental Services, also sees the series of accidents at Lanxess as constituting a problem: "I think there is a legitimate concern for air quality. We feel it is a serious enough issue that we're going to make sure the experts . . . determine if there were any health impacts." Kramer went on to say that it was unusual to have three accidents in a row such as those experienced at Lanxess and that it worried him.[15] Kramer officially notified the company that it had violated state air pollution laws in three incidents since October.[16]

On March 9, 2005, Lanxess officials reported an additional accidental release of 170 pounds of VOCs and four days later conceded that 99 more pounds of VOCs were released. Acrylonitrile, butadiene, and styrene are all VOCs, notes Breech, and could have been among the chemicals released on either of these two days. Furthermore, VOCs, when mixed with sunlight and heat, create ground-level smog that can cause lung irritation; they are considered a health hazard.[17]

### Working the Media

One of the first reporters to latch onto the Lanxess pollution story and stick with it was Hagit Limor, a journalist who had worked in Tampa, Florida, and Asheville, North Carolina, prior to moving to Cincinnati in

1994. With thirteen years under her belt as a reporter for WCPO's Channel 9 and five years as anchor of the I-Team, Limor knew enough about the area's industrial history that she could see that the series of accidents at Lanxess deserved in-depth, serial reports. Limor had known Lanxess in an earlier corporate incarnation as Monsanto, she explains. "Usually no one is around long enough as a reporter that they can see a pattern of violations," Limor observes, but her longevity on the beat allowed her to detect a problematic pattern with the Addyston plant.

Arguing forcefully with her producers that the story deserved major coverage, Limor put together a six-and-a-half-minute segment on the releases, which was promoted heavily before it aired, and then followed it up with a dozen shorter pieces as the story unfolded. This kept the Lanxess story front and center in the Cincinnati metropolitan area. Dan Klepal, a reporter for the *Cincinnati Enquirer*, also covered the story from the beginning.

"Most Americans expect that the government will protect them from chemical releases [such as those at Lanxess]," Limor said over lunch in Cincinnati. "But what emerged as I did these stories was that government officials do not lead on these issues but rather need to be led. It has been left up to grassroots groups and the media to shine a light on chemical pollution problems. As a result, a lot of corporations that are responsible for a lot of pollution are flying under the radar . . . and that leaves the population at risk."

This is hardly atypical. Across the nation, industrial plants and military bases irregularly emit dangerous levels of toxic chemicals into residential neighborhoods. These illegal emissions more often than not go unreported. State and federal regulators, overwhelmed by the number of sources of pollution they must monitor, are by necessity focused on only the most egregious releases. This leaves it up to grassroots groups and reporters to bring toxics problems to the attention of the public to bring pressure on polluters.

Health Impact

At the thirtieth anniversary of OCA, Richard Challis presented an award to Limor for the reporting she aired on pollution escaping from the Lanxess plant. Challis, who has lived for the past twenty-five years with his wife, Emily, in a house eleven miles from the plant across the river in the town of Erlanger, Kentucky, describes himself as exquisitely sensitive to butadiene, one of the chemicals used at the Lanxess plant. When butadiene is emitted by the plant and drifts across the river to his house, Challis experiences acute respiratory distress: "It is like a seizure. I'm rolling around on the floor gasping for breath unable to breathe." Emily Challis has kept a careful log of these episodes and has documented twelve thousand incidents over the past quarter-century. Over the past year, she has compared the log she keeps with similar odor logs kept by Addyston residents and found that they track closely.

Challis notes that these chemical sensitivity attacks don't take place on July 4, when the plant closes down, providing further evidence to him that Lanxess is the source of his problem. "I don't think I have patriotic allergies," he observes. The worst time for Challis is from midnight to 4:30 a.m., when he believes the plant emits the most butadiene. Unable to sleep during those hours, Challis has taken a job on the second shift at the airport so that he can sleep in the late morning. On a recent vacation to Europe, Challis found that his allergies disappeared. "It was wonderful to sleep through the night," he adds.

"There are four hundred people working at Lanxess, but it is not worth keeping those jobs if they are poisoning thousands of people," Challis says. "The rate of asthma in children in Addyston is astronomical, and the amount of cancer is horrible," he continues. Challis believes that the Lanxess plant is "poisoning the air." Even the trees are dying, he notes. Since the plastics plant began operations in 1952, the population of Addyston has plummeted from 1,600 to 982, he continues. Many people have gotten sick and moved out. Since Lanxess officials installed new valves at the plant, Challis has noticed some improvements in the air, "but do I think the plant will ever be a safe plant? No I do not. . . . People are dying from this slowly."

With putrid odors in the air and heavy media coverage of accidental releases from Lanxess, many Addyston residents began to voice their concerns about the impact on their health from this local source of pollution. Some began to speculate that cancer rates in town were elevated. Nancy Scott, a forty-eight-year resident of the town said that her mother, mother-in-law, brother, uncle, and grandfather—all Addyston residents— had cancer. "You go up and down these streets, and just about every house has experienced it. I don't want to see anybody lose their job at

Lanxess, but I don't think they have been honest with us," she told a reporter.[18]

Similar concerns were voiced by Sue Lloyd, a sixty-three-year-old breast cancer survivor who has lived with her husband in Addyston three blocks from the plant for forty years. "I've worried about it ever since the first time I smelled this odor," Lloyd was quoted as saying in a local paper. "It would burn my nostrils, and my eyes would tear, and my throat would get very raw from it. It would just take your breath away."[19] In enumerating those in Addyston who have cancer, Lloyd includes herself, her next-door neighbor, two people two doors down from her home, two residents in the house behind her, friends at the plant, and many others. "When you know so many people who have had it [cancer], it's just distressing," she observes.

## Elevated Cancer Risk

These anecdotal reports of a cancer cluster had no scientific standing until 2005, when a study conducted at the request of the Hamilton County General Health District and the Ohio Environmental Protection Agency (OEPA), found that "people who inhaled fumes for decades from the plant have a 50 percent greater risk of developing cancer" than the population at large. Among the chemicals of concern are acrylonitrile and 1, 3-butadiene, both of which are used by Lanxess and linked to cancer in humans.[20] Lung cancer and leukemia would be the most common forms of cancer caused by the chemicals released from Lanxess, notes Paul Koval, an air pollution toxicologist at OEPA.[21]

On May 25, 2006, over one hundred residents crowded into a VFW hall in Addyston were told that cancer rates in their community were 76 percent higher than expected in the general population. This estimate came out of an Ohio Department of Health Study that found that 55 residents in town were diagnosed with cancer from 1996 to 2003 compared with an expected rate of 31.2 cases. Lung cancers were four times higher than expected and mouth and colorectal cancers three to four times above normal. Health commissioner Tim Ingram described the Addyston cancer rate as troubling: "This study does not rule in Lanxess [as the cause of the additional cancers], and it does not rule it out." He called for follow-up studies.[22]

Officials at Lanxess deny that their plant is the cause of elevated levels of cancer in the community. Sandy Marshall, Lanxess plant manager, argues that concentrations of chemicals emitted from Lanxess are hundreds of times lower than those associated with cancer development and that although butadiene can cause cancer if inhaled continuously over a number of years, the study found none of the cancers associated with the chemical were "statistically significant."[23]

Reacting to community health concerns and community sampling data about the impact of emissions from Lanxess, the Hamilton County Department of Environmental Services installed air monitoring equipment on the roof of the Meredith Hitchens Elementary School located across the street from the plant.[24] The installation of the rooftop monitor came after citizen air monitoring revealed serious problems. On May 6, 2005, an air sample collected by local activist-resident Cheryl Siefert found 87 ppb of butadiene in her backyard. Ruth Breech described the reading as "extremely high" and urged the installation of more sophisticated monitors that continuously take samples. Lanxess officials questioned the accuracy of the sample.[25]

**Capturing Evidence**

Addyston residents also got help monitoring their air from Hilton Kelley, executive director of the Community In-Power Development Association (CIDA) from Port Arthur, Texas. One sample Kelley captured in Addyston on May 14, 2006, detected 219 ppb of butadiene in the air. Butadiene causes reproductive and developmental disorders, as well as eye, nose, and throat irritation. In high doses it can affect the central nervous system, and it has been shown to cause cancer in laboratory animals. Mike Kramer, permits and enforcement supervisor for the Hamilton County Department of Environmental Services, described that sample as well above the level that causes physical irritation. At 1,000 ppb, the chemical can cause nervous system damage and unconsciousness.[26]

Kramer faults inefficient flaring as possibly the source of a considerable amount of pollution that escapes the Lanxess plant. Flaring is a process by which chemicals are released up the stack and then ignited. This process is supposed to be 99 percent effective, but because the wind blows the gases around when they come out of the chimney, it can be as little as 20 percent effective. Kramer notes that monitors find spikes in the amount of pollutants detected after a dump-and-flare cycle.

Some residents near the plant are so concerned that they have taken on the job of doing their own air monitoring with air sampling buckets made from five-gallon plastic buckets, special plastic liners, and sealed lids with a small air pump to create a vacuum. These citizen air sampling devices cost about $150 each to produce, notes Denny Larson, director of Global Community Monitor, who introduced their use to the region. The buckets are in use in thirty communities around the country and a dozen other locations around the world, Larson explains. They have been useful in proving that air is polluted in fenceline communities adjacent to heavily polluting industries.[27] OCA was awarded a $7,500 grant to ensure that air samples taken by residents could be processed by an independent California laboratory. The county also issued some air sampling canisters to residents so they could test the air when they smelled suspect odors.

## School Closed

On December 6, 2005, Three Rivers School District officials closed the Meredith Hitchens Elementary School attended by 370 preschool to first-grade students. Their decision was made after hearing the results of OEPA's seven-month monitoring of air quality in Addyston across the street from the Lanxess plastics plant. "These air pollution levels of these two compounds [acrylonitrile and butadiene] on the school were higher than we consider acceptable for public health," OEPA toxicologist Paul Koval told Channel 9 News.[28] Koval went on to say that the concentrations of chemicals detected on the roof of the school posed a higher-than-normal risk of cancer for one out of every two hundred residents of Addyston over a period of thirty to seventy years. This rate is a far higher level of risk that the one in ten thousand level than the agency deems acceptable. OEPA also ordered Lanxess to reduce emissions of acrylonitrile and butadiene and carry out a planned $2.5 million environmental improvements scheme. "All along we have said whatever recommendations were made by the EPA, we would follow them," observed Rhonda Bohannon, superintendent of the Three Rivers School District. "Today they said they were concerned. We will definitely get our kids out of there." [29]

For Jennifer Janzen, a mother who pulled her children out of school because they complained of headaches, this was a moment of vindication. Janzen had tried to approach school officials about health concerns related to emissions from Lanxess but had been rebuffed. When she took her children out of school, they were listed as absentees, and some neighbors thought that she was just trying to stir up trouble, Ruth Breech recalls. But with the school being closed, Janzen suddenly appeared prescient.

The question of whether to close the school split residents in the communities surrounding the Lanxess plant. "I'm elated," said Sue Wullenweber, who lives near the plant and has a son at the school. "We need to think about the safety of our children. . . . I guess they have gotten the proof that our children aren't safe . . . [that] there's a true, real concern here, and we aren't crazy," she added.[30] "They should keep Hitchens closed," agreed Charity Hollin, whose six-year-old daughter, Madison, complained of headaches and stomach pains after beginning kindergarten at the school.[31]

But many Addyston residents were against the decision, arguing that it did not make sense to close the school while continuing to permit children and pregnant mothers to live in the same area in the town across from the plant. Although the school closed, the playground next to it remains open, and children continue to play there. "What sense does that make?" asks Lynn Bowman. Other residents are glad the school is closed, arguing that many of the standards for chemical exposure are for full-grown adults and that children with lower body weights and less well-developed immune systems are more susceptible to ill effects from chemical exposures. Complicating the school closure debate is the economic divide between modest-income residents of Addyston and their wealthier neighbors in the communities uphill, some of whom live in large, expensive homes overlooking the river. Some residents of Addyston suspect that the wealthier folks living in surrounding neighborhoods were using pollution problems from the plant as an excuse to ensure that their children would no longer be sent to the low-income school in Addyston.

**Industry Perspective**

On the other side of the fenceline, some employees at Lanxess argue that the plant does not constitute a threat to the health of its neighbors or workers and that they are making significant progress in reducing odors that come from the plant and accidental releases of chemicals.

Those who work at Lanxess see the plant as an integral part of the community. Industrialist Matthew Addy opened the pipe foundry in 1891, and a factory town grew up nearby to supply workers. The factory subsequently became a plastics plant and was purchased by Monsanto in 1951 and then by Bayer in 1996. It currently makes plastic pellets, which are sold to companies that melt and mold them into dashboards, telephone headsets, blenders, and refrigerator liners. The company has 410 employees and a $32 million payroll. In 2004 Bayer spun off Lanxess as a subsidiary.

Tim Bentner, a big-shouldered forty-three-year-old man wearing a yellow hard hat, has worked at Lanxess for twenty years, and his father and brother have worked there longer. He is convinced the plant is not causing illness in the community. Bentner grew up in Addyston, attended school there, and now lives in the nearby neighborhood of North Bend, where he can see the plant from his second-story window. After working at the plant for so many years and living next to it, he argues that if anyone had been made sick by the chemicals, it should have been him.

"If anything happens at the plant, we are seen as guilty before being proved innocent," he complains. The media have not done a good job of reporting the story about the plant, he continues. Reporters mention the high cancer rate but fail to point out that the cancers that are the highest are not ones associated with exposure to the chemicals used at the plant. They also fail to report that cigarette smoking among Addyston's residents is three times the national average. Bentner believes the operation of the plant is safe enough that he continues to work there and would urge his children to work there when they are older. He argues that only a very small minority of people in the surrounding communities are concerned about health effects from the plant.

He acknowledges that community members have every right to be angry about the three accidents. But the vast majority of 102 malfunctions reported the previous year are minor technical difficulties that the company is required by law to report but constitute no threat to the public. OCA, he adds, has misrepresented the facts, used scare tactics, and created an unwarranted sense of panic in town. However, Bentner concedes that OCA's campaign has been successful at challenging plant managers to do a better job, and it has hastened the purchase and installation of pollution control technologies which he sees as a good thing. "But they have left the community with bitter resentment about the plant and people no longer trust the facility" and that, he asserts, is unwarranted.

Duane Day, the Lanxess manager whose job is to track the company's environmental compliance with state and federal regulations, also has problems with OCA's campaign to clean up the plant. "The campaign took a negative approach, and they were not clear [about] what they wanted," he observes. First, they focused on odor problems, then dust, then water contamination, then they were worried about upsets (that is, accidents), and then they wanted a cancer study. The campaign never seemed to have clear priorities, and whenever the company began to respond, they would shift to another issue, he adds.

One of the biggest issues for residents was the odors from the plant. In response, Lanxess invested $300,000 to improve its sewer treatment plant and reduced the odors by 90 percent, Day points out. As for the three accidental releases that received so much media attention, Day says that although they were not a cause of health problems in the community, it was not acceptable for the company to argue that these malfunctions were just a cost of doing business. "We recognized that there was room for improvement," Day observes. As a result, the company committed $1.5 million for new equipment to improve the efficiency of flaring and spent $300,000 to replace some giant valves, he explains.

The company is also reaching out to the community and going door to door talking with residents. "A high percentage of people are okay with us. Only a few had negative comments," Days reports. It also has improved its complaint process, so when residents call to complain about an odor, Lanxess sends someone out to talk to the resident and attempt to capture information immediately in order to investigate further.

"I struggle with the accusations that Lanxess is making people sick," says Kay Rowland, age fifty-three, a Lanxess human resource coordinator who started work at the plant in 1980 at the age of twenty-six and lives a seven-minute drive from work in Miami Township. "I'm here at ground zero, and I have always felt safe working here," continues Rowland, who observes that many Lanxess employees have parents, grandparents, aunts, and uncles who have worked at the plant.

Rowland is not convinced that many of the cancers in Addyston should be attributed to chemical releases from the plant. As an example, she cites the case of her father-in-law, who died of cancer in 2002 and lived across the street from the plant. "He will be counted as one of the fifty-two cancer cases, but he both drank and smoked for years prior to his death, she points out. "I don't agree that it is all about Lanxess." Other factors may also explain the increased cancer rates, Rowland points out. A lot of people who live in town are in older houses where radon may be a problem, she observes. In addition, multiple other industrial operations such as Kaiser Chemicals and Cinergy are putting out pollutants.

Rowland thinks that the complaints about the plant have been "blown out of proportion." She agrees that sometimes the smell from the plant is annoying: "I don't like the sour egg smell, but we are getting rid of that." She is particularly incensed about the OCA campaign to change the management of the plant; it successfully mounted a petition drive to force into early retirement her former boss, plant manager Bill Ward. "He was a decent man and did what he could for the community," Rowland recalls tearfully. His retirement party was like the receiving line at a funeral, she recalls.

The new plant manager, Sandy Marshall, arrived in July 2005 and immediately tried to set a new tone with residents. His message to residents was that the company was going to make significant investments in equipment that would cause the plant to operate more efficiently and with fewer accidental releases. Having the plant be in compliance with regulatory standards was important, Marshall observed, but it was not enough. "We want to go beyond compliance and meet community expectations for the way we operate," he says. After all, it is really community expectations that drive whether a company has a right to operate a facility, he notes.

Asked about recent complaints about foul odors coming from the plant that are causing residents headaches, coughs, and tearing eyes, Marshall warns that improving a plant is a long-term process. "Are we at the holy grail yet? No," he concedes. But the trends are good. Releases of butadiene are down from 5 to 10 ppb to 1 to 2 ppb, and although reductions in releases of acrylonitrile are not as impressive, company managers are starting to focus on them more.

As for the possibility that releases from the plant cause cancer or other health problems in the community, Marshall is skeptical. The plant is meeting occupational standards and has a healthy workforce despite the fact that exposures to chemicals in the plant are likely to be one thousand times higher than in the surrounding residential community, he points out. And most of the cancers reported in the community are not ones associated with the chemicals used at the plant, he adds. Marshall is also "not convinced" by OEPA's risk assessment projections that point to a fifty-times-higher probability of contracting cancer among residents with long-term exposure to chemicals coming from the plant. And he was "disappointed" by the decision to close the school across from the plant.

The company will also challenge in court a state regulatory standard that was recently imposed requiring the plant to lower the level of butadiene in the air outside its fenceline to less than 1 ppb, a level lower than the "background" level that exists around the metropolitan area, Marshall continues. "They are doing administrative law setting," he asserts, without going through the normal steps involved in setting regulations.

On June 14, 2006, the U.S. EPA issued an eight-page Notice of Violation against Lanxess citing air pollution problems and leaks in the chemical piping system, and asking questions about wastewater discharge. Plant manager Marshall took exception to some of the conclusions drawn by the notice of violation but said plant workers had already fixed or were working on four of the six issues raised.[32]

### New Investments in Pollution Controls

The new Lanxess management team is beginning to make positive changes and has committed money to upgrading pollution controls, but whether they will carry through on all their commitments remains to be seen, says Sandy Buchanan, OCA executive director. "We are going to continue to track this very closely," she continues. Buchanan has been following the plant's environmental record ever since it was operated by Monsanto, when it already had a history of lax environmental management and accidental releases. "Everyone knew the plant was a problem even back then," she recalls.

"Our most significant contribution is that we took the problem that was festering in Addyston for fifty years, and we forced people to do something about it," she continues. As a result of OCA's high-profile good neighbor campaign that highlighted toxic releases from the plant, regulatory officials suddenly started discovering problems with the plant, the releases, the cancer rates, and the school. "Ruth Breech did an amazing job of helping neighbors organize and meet every week," Buchanan continues. She went into a highly charged environment in a small company town where the mayor was not happy with an outside group raising issues about air quality, and she did not back down, Buchanan says with undisguised pride. Through canvassing and walking-and-talking tours through Addyston, Breech found people willing to speak out and press Lanxess to clean up. "As a result I think we had a significant impact on the way the company operates," Buchanan concludes.

Whether Lanxess officials will be able to significantly reduce everyday and accidental releases from the plant remains an open question. What seems sure, however, is that the media and environmental groups will continue to follow this story closely and that the residents of Addyston, now informed about the dangers of some of the chemicals being handled next door, will continue to watch their neighborhood industry for releases and demand improvements.

## State Officials Play Critical Role

Ohio state public health officials and environmental regulators played an important role in responding to illness and odor complaints from residents who lived near the Lanxess plant. Too frequently residents plagued by pollution in towns such as Addyston, get no relief from the county or state departments of health when they complain of being made ill by emissions from a neighboring factory. Often citizen groups must conduct their own unfunded and informal health surveys after they notice what appears to be an excessive number of cancers.

Usually these anecdotal stories of cancer clusters and elevated levels of respiratory disease are dismissed by epidemiologists and health professionals as unscientific. Only occasionally are these resident health surveys followed up with professionally administered health studies. The description of what happened in Addyston is one of the rare instances in which state officials mobilized to answer questions raised by residents. County officials monitored air quality on the roof of a school across the road from the plant and found unsafe levels of dangerous pollutants. State public health officials subsequently conducted an epidemiological study that revealed an excess of cancer cases in the community.

In the future, to better detect fenceline toxics problems before they become widespread, state officials need to design a system for collecting targeted morbidity and mortality data in residential areas adjacent to heavily polluting industries. They also need to convince doctors who treat patients who live near these hot spots of pollution to report to state public health officials any symptoms they see that could be caused by pollution. Since this system is not in place, it is often left to local residents, such as those in Addyston, to sound the alarm about dangerous environmental conditions. Most likely the public health investigation in Addyston would not have been undertaken without the intervention of Ruth Breech at OCA and local residents who raised the alarm about dangerous releases from the plant and convinced reporters to put the news before the public.

Following the campaign she helped spearhead in Addyston, Breech began to work in Marietta, another Ohio River Valley town on the other (eastern) side of state located across the border from Williamstown and Vienna, West Virginia. As we will see in the next chapter, residents of Marietta were inhaling emissions from a steel-hardening plant that emitted tons of manganese-laden dust, a heavy metal that can cause subtle neurological damage. Because the research into and standards about manganese exposures are less well established than the health impact of chemicals emitted by the plastics plant in Addyston, Breech and local residents in Marietta faced not just an organizing task but also a public education job in order to convince company officials to invest millions of dollars in pollution control equipment.

# 6

## Marietta, Ohio: Steel-Hardening Plant Spews Tons of Manganese into River Valley Town's Air

Caroline Beidler did not suddenly decide to organize a local antipollution campaign. Her leadership role, which she was to assume later, crept up on her. Beidler, who works at an advertising agency, and her husband, Keith Bailey, a carpenter, built their "dream home," one board at a time, at the end of a dirt road in the wooded hills outside Marietta, Ohio. With a view from their windows of uninterrupted forest, they were convinced they had found their own private paradise.

What they did not realize at the time was that their edenic homestead was located just four miles (as the crow flies) from Eramet Marietta, Inc., a sprawling, French-owned, ferroalloy plant. Eramet (which uses manganese, cadmium, and lead, among other feedstocks, to strengthen steel and purify chromium) releases tons of heavy metal dust into the air. It is one of the county's top polluters.

It took awhile before Beidler and Bailey acknowledged to each other that they had a problem. As Beidler recalls it, they turned to each other one night and said, "What's that stink?" The industrial stench was so foul that it sometimes woke them up in the middle of the night, Bailey reports.[1] Beidler began to keep what she came to call a "stink diary," in which she logged odor events reported by a network of residents from nearby communities who called themselves the "Stink Club." At first they were not sure from where the odor came, but it was not long before they tracked it to the Eramet Marietta steel-hardening plant.

Beidler describes the odor as an ammonia-like smell mixed with heavy metals that leaves a burned metallic taste in the back of the throat and a sense that one's teeth have become "gritty." When there are heavy releases from the plant, often at night, neighbors near Eramet call Beidler and tell her that an odor just entered their home and that she could expect its arrival at her house in about an hour. And sure enough, the stink would roll down the valley to her house.

"It is an awful industrial odor that lingers into the early morning hours and then dissipates," explains Beidler, who says that because of the topography of the area, her home is in a kind of alley that channels pollution from the plant to her front door. Not only do the chemicals wake her up from a sound sleep, they also cause watery eyes, a cough, throat irritation, nasal congestion, nosebleeds, nausea, and vomiting. From 1998 to 2000, Beidler logged an average of eighty-seven stink episodes a year.

A few more recent excerpts from Beidler's stink diary, which she began in 1998, give a sense of what local residents endure:

• "Wednesday, November 7, 2006—10:30 p.m. Coming back from Athens [nearby town] in front of Eramet, thick, toxic fog.
• Monday, November 19, 2006—walking dog, could see haze and smell odor.
• Tuesday, November 20, 2006—terrible headache all day. Warm night, went to open window at 9 p.m. AWFUL, thick smell. Haze over valley all day."[2]

An entry from October 30, 2006, sounds particularly desperate: "Hi Neighbors. It is 4:30 a.m. in Pinehurst [a township near Marietta]. I am awake because Eramet's industrial odors are strong enough to be in our home. Keep in mind we are at least four miles from Eramet. Yes, our doors and windows are closed. We have headaches and sore throats. I have called the Ohio EPA and I am waiting for the duty officer to call me back. I'd much rather be sleeping. . . . At times like this do we think of moving far, far away? Most definitely! Sincerely, Caroline in Stinkville."[3]

Initially Beidler was reluctant to call regulatory officials out of fear that if she lodged a complaint with the Ohio Environmental Protection Agency (OEPA), she would "get the plant in trouble." She later saw this concern as hopelessly naive as numerous calls failed to provide regulatory relief.

### Large Volumes of Toxic Heavy Metals Released

With little help coming from state regulatory officials, Beidler and Bailey purchased an air monitoring device that detected, among other substances, tetrachloroethylene, "a chemical that can cause dizziness, headaches, nausea, unconsciousness, and even death." This chemical was not among the long list of chemicals that Eramet reported having released. However, in 2004, Eramet did emit 15,000 pounds of chromium compounds into the air and 75,000 pounds into the river and 500,000 pounds of airborne manganese, a known neurotoxin.[4]

Eramet officials say they are making efforts to reduce emissions and odors. "We're working to do whatever we can to eliminate our contribution [to the problem,]" says Jeff McKinney environmental manager at the Eramet plant. Total releases of toxic chemicals by Eramet reported to federal officials were radically cut from about 12 million pounds when the company was purchased in 2000 to about 6 million pounds of TRI releases in 2004.[5] Eramet is not the only local contributor to air pollution. Emissions from a large number of industries collect in the river valley on foggy and windless days, he continues. "They are getting pollutants from everybody. But because we are last in the chain, they are blaming it entirely on us." Eramet has installed a variety of pollution control equipment including baglike filters, explains McKinney, who adds that the plant is meeting environmental standards.[6]

**Getting Organized**

Unconvinced that Eramet was doing all that it could do to suppress emissions, Beidler and Bailey gradually transformed the network of neighbors who provided entries for the stink diary into the Stink Club, then the Odor Task Force, and finally they became founding members of Neighbors for Clean Air, a grassroots environmental group focused on cleaning up the air in the region of river valley towns around Marietta. Over time, this group began to collect information about air quality in their region and make their network of members aware of key regulatory developments, scientific studies, health studies, and emissions at Eramet.

For example, in mid-September, 2003, the EPA announced that Eramet's furnaces had emitted "more particulate matter during a June stack test than is allowed by EPA regulation." In May 2004, the Agency for Toxic Substances and Disease Registry (ATSDR) monitoring study of nine industrial contaminants from plants in the Ohio River Valley indicated that "manganese levels warrant additional monitoring." Another discouraging trend revealed that although toxic releases were declining statewide, in Washington County, where Marietta is located, emissions actually increased by 500,000 pounds between 2003 and 2004.

Then in December 2005 came news that galvanized many local residents into joining Neighbors for Clean Air. A report by David Pace of the Associated Press listed Eramet as the top factory nationwide "whose emissions created the most potential health risk for residents in the surrounding community." Furthermore, Washington County was ranked number one for the "highest health risk from industrial pollution in 2000."[7] These revelations stirred both concern and the beginnings of an activist response in the communities surrounding the Eramet plant. One of the families brought into the struggle to clean up the air was the Kuhl household in Marietta.[8]

## American Dream Evaporates

"We thought we had the American dream," says Lesley Kuhl, who since 2002 has lived with her husband and two young children on a quiet, leafy street in Marietta. Everything seemed to be in place to live the good life: she found work in town as a lawyer, and her husband, Dennis Kuhl, was a physics professor at Marietta College and could walk to work. They joined a church they attend regularly and made many friends in the community. The geographic location of Marietta was also fortuitous for the Kuhl family because their children could easily visit both sets of grandparents, who lived not far away. There were other activities as well: when Lesley Kuhl was not busy with her work as an attorney, she played harp at weddings and other social gatherings.

"We really loved living here. But then the American dream evaporated when we found out there was an invisible threat to our children's health in the air," says Kuhl, referring to heavy metal pollution coming from Eramet. Before moving to Marietta, Kuhl had not focused much on environmental issues. It wasn't until December 2005 that she read an article in the local newspaper about elevated levels of heavy metals in the air and their possible impact on the development of the brains of very young children. This information set off some serious alarm bells with Kuhl, whose children had suffered numerous sinus infections that had to be treated with antibiotics, and one of whom was diagnosed with a developmental disorder.

Trained as a lawyer, Kuhl wanted facts: What was in the air? Was the air dangerous to her children? And what was the government doing about it? She also wanted to know if she should move her family immediately to avoid subtle losses of intelligence and Parkinson's-related illnesses. To get some answers, Kuhl and about sixty other concerned residents attended a presentation by epidemiologist Stephanie Davis from ATSDR in mid-September 2006. One resident asked the question that was on Kuhl's mind: "If you had family living here, what would you say to them?"

Davis replied that despite the fact that "the highest reported air releases of manganese in the U.S. are in this community," she would "have to come down on the nonreactionary side" until there was solid evidence of a need to move. "Manganese is a neurotoxicant, and the main concern is its effect on the nervous system," Davis continued. "Health studies have showed that older people may have mood and movement problems from exposure," she added. The agency would expand its monitoring program and was considering conducting a health study that might take several years to complete. "We think this is a step in the right direction. It might be too slow for some people, but we are making some progress," she concluded. [9]

Kuhl was outraged: "These are my children: I don't have three years," she says. Her husband sounded similarly upset in a letter he wrote to the editor of a local newspaper—one of many such letters that were to become an outlet for residents to express their dissatisfaction with the quality of air in their community. Noting that Eramet was the largest contributor to air pollution problems in the county, Dennis Kuhl asked why officials at the plant who live in the area were not taking decisive steps to solve the problem. "Where is the compassion for our children? Where is the shame," he asked?[10]

The Kuhls faced an awful choice: Should they uproot their family because of an invisible health threat or stick it out? Not having hard evidence on which to base a decision was frustrating, says Lesley Kuhl, who admits that she sometimes regretted moving to town. Her husband was also concerned: "If you knew that staying here would reduce your children's IQ by five points, wouldn't you move?" he asks. But relocating without knowing that it was necessary seemed foolish. The Kuhls were in the early stages of their professional careers, and both had good jobs: "Professionally it was not a good time to move," she observes.

There was also the sense that if they relocated, they might unwittingly move from one area with a yet unproven health threat to another neighborhood with undisclosed chemical contaminants. "We could give up one problem here for another wherever we move. We just don't have enough evidence," she laments. Weighing in on the side of not moving was another consideration. Staying in Marietta and fully investigating the heavy metal air pollution problem and doing whatever they could to improve the situation "might serve some purpose higher than ourselves," she notes.

Kuhl's sense that she might be moving from one area with a pollution problem to another with a different problem is not at all unreasonable. As two academics interested in environmental health issues wrote: "In thousands of communities across the United States, billions of pounds of highly toxic chemicals including mercury, dioxin, polychlorinated biphenyls, arsenic, lead, and heavy metals such as chromium, have been dumped in the midst of unsuspecting neighborhoods. These sites poison the land, contaminate drinking water, and potentially cause cancer, birth defects, nerve and liver damage, and other illnesses."[11]

Somewhat reluctantly, Kuhl decided she had to take action: "I'd rather be playing tennis than be stuck doing all this research and writing letters trying to get these people to run their plant responsibly." But she did not feel that she had that luxury and started attending meetings of Neighbors for Clean Air.

### Working the Media

In a letter she wrote to the *Marietta Times*, Kuhl proved how forceful a spokeswoman she could be for parents in Marietta. Her children were under six years old, were in the midst of their neurological development years, and were exposed to manganese levels in the air that were ten to twenty-five times higher than recommended by state regulators, Kuhl observed. Then she unloaded on two statesmen in Columbus, the state capital, who had said that "we all breathe the same air and we all have the same concern for its quality." No, we don't, she wrote: "Do their children live within a five-mile radius of one of the largest emitters of toxic air pollutants in the entire country, as mine do? When was the last time either of these two men took their 2-year old child to have blood drawn to determine his manganese level? When was the last time these gentlemen drove their toddler to a neurotoxicologist three hours away to be examined for an alert level manganese exposure? My family had to do both last year."[12]

As time passed and little was done to control emissions from Eramet, Kuhl became more outspoken in the letters she wrote to the local newspaper. To get her point across, she personalized the issue: "Imagine you invited me to a backyard gathering. When I arrived, the first thing I did was spray neurotoxic chemicals into the air near your children, elderly parents, and friends. You'd certainly tell me to stop, right? What if I refused? You'd throw me out and notify the authorities of my actions, and rightfully so. Would you want to be my friend or invite me back? Would you like me to start a company in your neighborhood? Would you come to work for me, so I could shower you with chemicals in the workplace?"[13]

The Kuhls did not neatly fit the demographic profile of stereotypical grassroots, antitoxic activists, and the clean-the-air campaign they joined took them somewhat outside their circle of friends and acquaintances. "We are lifelong Republicans, and we are both socially conservative," she explains. Signs of the Kuhls' patriotism and deeply held religious beliefs are not hard to spot. The outside of their home is decked with an American flag and red, white, and blue bunting; inside, their two- and five-year-old children play next to a sign that reads: "I'm Growing Up Like Jesus." By contrast, many of the residents with whom the Kuhls ended up working to reduce Eramet's emissions were socially liberal. But the pressing nature of the pollution problem in Marietta transcended politics, "and the group did a good job of keeping the discussion nonpartisan," she observes. "Each person brought their own skills to the table," says Kuhl, whose training as a lawyer was invaluable.

### Health Impact

Another Marietta resident who brought relevant skills to the table was Dick Wittberg, who heads the Mid-Ohio Valley Health Department. Wittberg began to question whether the hundreds of thousands of pounds of manganese dust emitted by Eramet, which rained down on Marietta and surrounding communities, might be causing neurological deficits in local children.

Concerned that children in Marietta might be at risk of being damaged neurologically by the heavy metal emissions, Wittberg carried out a pilot study in the late 1990s that compared the ability of children in Marietta to perform physical tasks and answer academic questions. He then compared the results from Marietta with a control sample of children from a similar-sized town in Athens, Ohio, located forty-five miles away. Wittberg "gave a battery of 13 tests to fourth-graders in both cities. The children had been matched for age, sex, and parental education. The tests measured such things as educational proficiency, balance, visual contrast sensitivity, and short-term memory."[14] What he found was disturbing: across the board: the Marietta youngsters scored significantly lower on the tests than did those from Athens. Although Wittberg is careful to say that these preliminary results do not prove that manganese is causing deficits among children in Marietta—there are other possible explanations—he does think that the evidence he gathered is disturbing enough to warrant that more comprehensive studies be conducted. "In my opinion, it [his study] points to some neurological differences and one has to suspect manganese. Nobody knows, for kids, how much [exposure] is too much."[15]

"There aren't people dropping dead from this [manganese exposure]," Wittberg explains. "Manganese exposure can rob potential. What if a few points are shaved off the IQ of our children? All of a sudden our brilliant children are just smart, and all our smart kids are just average. . . . It's not something you would notice unless you were looking. I believe it is the whole lives of kids that are affected. I don't think that damage can be undone." He continues, "I'm not out to close Eramet. I hate to see plants close. But if they are hurting people, I do believe we have to do something to control emissions and the technology exists to do that."[16]

Eramet's environmental manager, Jeff McKinney, is skeptical of Wittberg's results. Although it is not clear if a final report on Wittberg's study is available, he writes, "It is suspected that the study was not performed using the proper controls, yielding results that are random and not statistically relevant." Ohio Scorecard, which ranks schools by aptitude and proficiency, shows that Marietta students rank either average or slightly better than the state average, he points out.[17]

## Capturing Evidence

Picking up where Wittberg left off, Erin Hayes, assistant professor of environmental health at the University of Cincinnati, is analyzing the blood of one hundred residents who have lived for at least five years within a ten-mile radius of Eramet. The blood samples will be analyzed for manganese, chromium, and lead, all three of which are emitted by Eramet.

"The problem is that during the refining process, manganese particles are released into the air," explains Hayes, who received a grant of $2.6 million from the National Institute of Environmental Health Sciences.[18] "When we breathe them [manganese particles] in they can travel directly to the brain and potentially cause neurological and behavioral disorders. We don't know at what levels these effects occur, and we are specifically interested in studying the effects of these metals on infants and children." Hayes will direct researchers to recruit an additional fifty participants for a "postural sway test" to determine their ability to maintain balance.[19] This test is relevant because excessive exposure to manganese can cause symptoms "similar to Parkinson Disease with movement and motor disorders, tremors, mood swings, depression, anxiety," she explains.[20]

In her preliminary assessment of manganese exposure levels in Marietta, Hayes is using swipe samples from rooftops, personal air monitors, and sway tests to delineate the areas most heavily affected. In addition to blood and hair samples, the study will examine dust, soil, and paint retrieved from the homes of participants. Hayes is also collecting "shed teeth" ("baby teeth") from children who lived in the area to examine for their heavy metal content. Thus far, Haynes has found that residents averaged 9.1 micrograms of manganese per liter of blood, a level significantly higher than a population near Quebec, Canada, that experienced a variety of negative health effects, including "shakes, tremors, loss of balance and more severe problems" with an average level of 7.3 micrograms of manganese. Hayes makes clear that a safe level of manganese in the system has yet to be determined.[21]

Beidler, a founding member of Neighbors for Clean Air, welcomed Haynes's interest in the potential health effects of Eramet's manganese emissions on the health of local residents. "The citizens of Washington County, Ohio and Wood County, West Virginia, have been exposed to airborne manganese for 50 years. The perception is that the particulate emissions we see coming from Eramet, the soot we wash off our homes and the manganese-laden air we breathe, is harming our health. Dr. Haynes's research adds legitimacy to our campaign to convince Eramet to modernize their equipment and lower their emissions," Beidler told a local reporter.[22]

## Intense Mid-Ohio River Valley Pollution

If manganese pollution does turn out to be harming the development of children in Marietta, it will not be difficult to figure out where most of the heavy metal dust came from. Eramet alone released 550,000 pounds of manganese in 2000. To get a more accurate picture of airborne pollution from the plant, ATSDR officials erected stationary air monitoring equipment within a five-mile radius of Eramet. Preliminary readings by these monitors registered airborne manganese at levels two to sixty times higher than the U.S. EPA health standard.[23] The EPA standard, established in 1993, set its health guideline for manganese at 0.05 micrograms per cubic meter of air. Local monitoring consistently found more than that in Marietta's air.[24] Although the initial study "found arsenic and manganese in the air consistently exceeded levels that scientists believe harm health," it would be premature to say whether the pollution constituted a health hazard, said Michelle Colledge, an ATSDR spokesperson.[25]

Some residents, however, had reached their own conclusions about the health impact of what some called the "Mid-Ohio Valley Crud." "Since I moved to Marietta in 2000 I have been diagnosed with asthma. I recently became a grandmother. I do not want my grandson exposed to the toxic substances in our air here in Marietta. . . . A co-worker of mine here in Marietta has two infants under the age of 2, both of whom have bronchial asthma. Our local statistics for respiratory illness and cancer are the highest in percentiles measured against other communities nationally. Our neighbors are dying from our local environment," wrote resident Ellyn Burnes in a letter to the editor.[26]

Although many local residents viewed the ATSDR data with alarm, Eramet officials were quick to point out that their company is not the only industry contributing to air pollution problems in the Mid-Ohio River Valley. An Associated Press report, which used EPA and census data, found that one-tenth of the total national-long-term health risk from industrial air pollution was concentrated in Ohio, particularly in the Ohio River corridor. This makes it hardly surprising that the nine neighborhoods around Marietta and across the river in Wood County, West Virginia, "rank among the worst 100 nationally for health risk from factory emissions."[27]

Tom Hockenbrocht, a jeweler from Marietta whose wife recently died of pancreatic cancer and who has been diagnosed with acute myeloid leukemia, says that although he loves the town of Marietta, had he known about the air pollution problem, he would have chosen another place to live. One day while Hockenbrocht was learning to fly a single-engine airplane, he noticed a yellowish haze hanging over Marietta and the adjacent river valley towns. When he asked his flying instructor about it, he was told that the chemical-laden mist was almost always over the town except on days when there was a strong wind. "You don't notice it so much when you're on the ground in the midst of it," the instructor added.[28]

### Neighbors for Clean Air

By February 2006, Beidler, Kuhl, and other concerned residents were beginning to meet regularly. At a meeting on February 20, some thirty residents gathered to complain about "extreme odors" from Eramet. "Almost everyone sounded fed up with the company, the Ohio EPA, the Ohio Department of Health and wanted to do something about the problem now. All were disgusted with the county's ranking worst in the nation for toxic air and saw it as hurting the local economy," Beidler wrote.[29]

Help for the residents in Eramet's airshed arrived on March 8, 2006, when Ohio Citizen Action (OCA), the largest environmental group in the state, launched a Good Neighbor Campaign, aimed at pressuring Eramet officials into reducing emissions. Ruth Breech, the southern regional director of OCA, was assigned to provide organizing support to the growing grassroots movement in Marietta that was protesting Eramet's emissions. Breech, a young but experienced toxics activist, had recently emerged from a successful campaign to organize residents along the fenceline with a plastics plant in Addyston, Ohio. The campaign had resulted in the company's agreeing to upgrade its pollution control equipment and reduce its emissions. This matched neatly what residents like Beidler and Kuhl in Marietta had as their goals.

In Marietta, Breech met with residents who were concerned about air quality and began to provide them with organizational resources. She used a variety of strategies to put pressure on Eramet to clean up. She started by assembling a team of young researchers in Cincinnati to dig out the facts about Eramet's history of emissions. She encouraged local journalists to write about the fact that residents were organizing around the Eramet pollution problem and provided them with facts about Eramet's pollution record that she found in EPA's Toxic Release Inventory.

As a result of these efforts, the news was soon broadcast that Eramet was annually dumping 6 million pounds of toxins into the environment, including 4 million pounds of manganese, some 312,000 pounds of which was spewed into the air.[30] Furthermore, there had been fifty-three known accidents and malfunctions at the plant since Eramet purchased the company in 2000. As a consequence, a community of some twenty-five thousand had been exposed to the plant's odors and airborne manganese, a known neurotoxin, since the facility opened in 1951.[31]

## Citizen Audit

As part of the Good Neighbor Campaign, Breech pulled together the densely researched, twenty-six-page "Citizen Audit of Eramet Marietta," which summarized information about releases from the plant, regulatory actions, monitoring studies, and a history of citizen activism. Among other revelations, the report exposed the fact that the Eramet facility in Marietta, in addition to being one of the top emitters of air toxics in the nation, released ten times more manganese-laden dust emissions than two comparably sized plants run by Eramet in Norway. This left plant managers to fend off criticism that they were exposing Americans to heavier loads of pollution than they were Norwegians. Residents in Marietta began to ask why the protection of the health of residents of Ohio was considered less important than that of European citizens.[32]

Breech made frequent trips to Marietta with OCA teams of outreach workers who went door-to-door asking local residents if they found problems with air quality and explaining to them their Good Neighbor Campaign. During one such foray in late August 2006, canvassers signed up 246 residents in Vienna, West Virginia, across the river from Marietta.

Breech and her team took swipe samples from the siding of homes in the area, which were found to contain both manganese and chromium, and they passed out pollution logs so that residents could record visible pollution and odor events. While driving past Eramet one afternoon, Breech observed "a large orange cloud rising from the plant's smokestacks. The orange dust lasted for ten minutes, and then dissipated into the air. There was a haze around the plant all weekend."

## Letter Writing Campaign

Breech mobilized local residents (as well as those from communities around Marietta and other parts of the state) to join a letter writing campaign. By the end of the campaign, 57,624 letters from OCA members statewide were sent to the managers and owners of the Eramet Marietta plant, including 6,399 letters to French president Nicolas Sarkozy urging him to direct Eramet to clean up its operations. (The French government is Eramet's second largest shareholder.)[33] Many other letters were addressed to the Eramet plant manager, Frank Bjorklund; and to the CEO of the company in Paris. One of these letters read as follows:

Dear Mr. Bjorklund:

I am writing this letter in regards to the pollution your factory is putting into out air. It sickens me to know that a company in the United States cares so little for the people who live here and their well being that they are willing to put so much pollution into the air when they obviously have the means to make our air better. Knowing that you have the technology in Norway to emit only 15% of the pollution you emit here in Marietta is disgusting.

We teach our children right from wrong, to treat others with respect, and to clean up after themselves. Our children, it is sad to say, know better than a grown man and the company he runs. Let's hope you can learn from them.

Be responsible for the mess that your company has made and do what is right. Eramet can and must reduce its impact on public health and the environment, commit to significant reductions of manganese air releases and odors. Citizens in the mid-Ohio Valley deserve to breathe free. Run your plant to the same standards as your plants in Norway.

Do what is right and responsible.

Andrea Richardson

Officials at Eramet were understandably discomfited by this letter writing campaign. Commenting on it, Jeff McKinney wrote that "it might have caused irreparable damage to the reputation of our facility and our community statewide. The misinformation spread by OCA in distant communities borders on libel."[34] He later added, "I have been frustrated and extremely overwhelmed by the amount of negative press and public opinion surrounding the Eramet facility in recent months." Attempts to answer these criticisms have "resulted in the publishing of severe misquotes or statements taken out of context."[35]

Marietta was by no means the only town affected by pollution from the Eramet plant. Vicki Dils, who owns a farm and stable across the river from Marietta in the town of Boaz, West Virginia, is outraged at the poor quality of the air. The air quality is sometimes so bad "that it is like the death of spring. How can someone kill spring?" she asks. Plucking a fruit from one of her apple trees, Dill showed a visiting team of documentary makers that it was covered with a gray-black, heavy metal dust that you had to work at with your fingers before it rubbed off.[36] "Everyone is hush-hush. We're not supposed to mention the real problem in our area. I think we would all agree we need to be able to breathe air, that life-giving force, without which we could not live. . . . All you have to do is go outside and breathe when the wind blows the poisonous stuff your unlucky direction," Dils wrote.[37]

Dils was not alone in thinking that industrial air contaminants were having a negative effect on the health of the local economy. Dennis Kuhl reports that several of his colleagues from Marietta College had left the area for a variety of reasons, including poor air quality. The bad air is keeping other cleaner-technology companies from moving in, he believes. Caroline Beidler's mother, Florence Beidler, agreed: "There are people who are not moving to Marietta because of air quality, and there are people who are looking to get out of here because of it."

While members of Neighbors for Clean air were highly critical of practices at Eramet that led to excessive releases of manganese, the aim of the group was not to close the plant. "The goal is simply to ask the Eramet facility in Marietta to use the best technology available to them to reduce air emissions in the Mid-Ohio Valley. . . . It's just that simple. What we DO NOT want to do is to lose Eramet as an employer in our community. They are a very large employer and a large taxpayer," observes John Whistler, a member of Neighbors for Clean Air. "It's in their [Eramet's] interest and our interest and everyone's interest to clean up," concludes Patrick Stewart, a member of NCA's monitoring committee.

**Industry Perspective**

Clearly stung by the campaign that Neighbors for Clean Air and OCA were conducting, officials at Eramet pushed back with a publication addressed to their employees: "For the Record." In it Jeff McKinley, environmental manager at Eramet, accused OCA of having a "tradition of manipulating, misinterpreting, and manufacturing data to create undue concern in affected communities, thereby gaining funding for their so-called 'campaigns' by any means necessary."[38]

Specifically McKinney found fault with OCA's contention that manganese, chromium, and cadmium found in Marietta's air by government monitors poses "a threat to residents close to the Eramet facility." This is not accurate, McKinney argues. The ATSDR report found that only manganese posed a potential health threat to residents near his facilities. Furthermore, the "actual manganese exposure data from this monitoring has not shown a significant health risk exists." Manganism (manganese poisoning) occurs only after "prolonged, repeated exposure at high concentrations," McKinney explains. It is not proven that there is "a direct correlation between elevated blood manganese levels and increased risk of disease or neurological damage. Nor have there been any known cases of manganism or manganese-related neurological damage among Eramet employees at the Marietta facility."

Second, OCA's contention that the Eramet Marietta facility is emitting ten times as much pollution as similarly sized facilities in Norway is inaccurate, says McKinney, who argues that "parallel comparison is extremely difficult to make, as reporting requirements and accepted methods of emissions calculation and estimation vary greatly from country to country." The Eramet facility is using state-of-the-art air pollution-control technologies to reduce manganese emissions. In total, Eramet spends $5 million a year "operating and maintaining pollution abatement systems that capture the vast majority of emissions and prevent them from leaving the facility property. Moreover the facility operates in accordance with stringent federal permits that limit the discharge of pollutants to the air, water and land."[39]

These explanations, about Eramet's efforts to control emissions from its vast industrial works, while no doubt convincing to a manager deeply involved in the mechanics of production, were unconvincing to some local residents. One of them, Joan Dearth, wrote that Eramet's cleanup efforts did not "ease the pain of families who have lost loved ones from cancer or other illnesses most likely caused by toxic emissions. . . . Who will speak up against this atrocity if we don't choose to do it ourselves? We should all be outraged because of the air that our children and grandchildren breathe here in Washington County is simply unacceptable."

## Inadequate Regulatory Response

Eramet officials were not the only ones on the receiving end of OCA's critique. The environmental group also wrote that the Ohio EPA's files on Eramet were both incomplete and disorganized and that the regulatory agency did not have a large enough staff to adequately oversee the company's operations. In response to this public rebuke, Ohio EPA director Joseph P. Koncelik wrote Breech a letter in July 2006 pointing out that the agency had slapped Eramet with a $3.25 million fine on a water enforcement case and was working with the ATSDR on air monitoring of manganese emissions and their possible health effects. Koncelik noted that unless the ATSDR monitoring stations showed that Eramet's manganese emissions posed an unacceptable health risk, there was little that his agency could do "to impose more restrictive requirements than those established by state and federal regulations." As for the odor problems, Koncelic noted that the many complaints his agency was receiving were most likely due to ammonia releases from the plant. Here again, his agency was constrained by the fact that while "humans can smell ammonia at concentrations as low as five parts per million, ammonia levels would need to be much higher to negatively impact human health." Between the lines Koncelic was telling Breech that state officials were already at the limit of what they could do to rein in emissions from the facility.[40]

## New Investment in Pollution Controls

After a twenty-eight-month Good Neighbor Campaign aimed at pressuring Eramet to invest in additional pollution control measures to suppress manganese emissions, local activists were looking forward to marching in the local Washington County Labor Day Parade with a banner saying

"Eramet—Let's Clean the Air." The previous year they had been barred from participating in the march.

For months negotiations had been dragging on among Eramet officials, state regulators, and representatives of Neighbors for Clean Air and OCA. Then came the breakthrough: the day of the parade, September 1, 2008, Eramet purchased a full-page ad in local newspapers announcing that it was prepared to spend $150 million for upgrades over a five-year period. The plan called for a state-of-the-art furnace as part of a "comprehensive transformation" that would lead to "drastic reductions in airborne manganese and odors." While the scope of the envisioned retrofit remained unspecified, planners were considering improving the plant's environmental and operational function by further developing its refining technology and possibly restructuring the overall layout of the plant. "All residents of the Mid-Ohio Valley deserve to live and raise their families in a dynamic community with rewarding employment opportunities, where environmental stewardship is a top priority, and where industry and residents can work together to solve problems and effect positive change. We intend to do our part. . . . The Time Is Now," wrote Frank Bjorklund, CEO of Eramet Marietta.

"Wow. We're so glad to see this," said Caroline Beidler.[41] Greeting the news as an important victory for citizen activism, OCA announced an end to the Good Neighbor Campaign and thanked the thousands of residents and people around the state who had written letters and e-mails, displayed yard signs, made donations, attended meetings, marched with signs, and generally held the feet of Eramet officials to the fire.

In their successful drive to convince Eramet owners to make a substantial investment in improved production facilities and pollution control technologies, grassroots activists in Marietta had support from a large state environmental organization that helped them ramp up their campaign. The state agency also placed pressure on the company to clean up. In addition, several health studies and monitoring programs provided data that strengthened the case for reform. Unfortunately this kind of felicitous convergence of a wide variety of groups working to improve environmental conditions and reduce the impact of industrial emissions does not always occur in communities afflicted with environmentally induced illness.

The next chapter turns to the tiny African American hamlet of Talle-vast, Florida, where residents were alarmed to learn that the shallow well water they had been drinking for decades was contaminated with trichlo-roethylene and other dangerous chemicals disposed of on the grounds of a nearby machine shop and weapons plant. This is not a story with a tidy ending: a cleanup is on going, a lawsuit is in process, and residents have yet to be compensated for damage done to their health or relocated to a safer area. The chronicle of events in Tallevast makes clear that winning local environmental justice campaigns often requires a protracted struggle, and the outcome is never certain. In fenceline communities such as Tallevast, residents do not have the sense that the government has adequately protected their health or that justice will be done. If justice is served—and they are compensated for damage done to their health, the quality of their lives, and the lowering of their property values—it will be a result of the determined organizing and the persistence of two local leaders: Laura Ward and Wanda Washington.

# III

## Contaminated Water

# 7

## Tallevast, Florida: Rural Residents Live Atop Groundwater Contaminated by High-Tech Weapons Company

On a September morning in 2003, a drilling crew pulled up onto Laura Ward's lawn and started boring a hole. "Why are they driving on my lawn?" Ward wondered as she sat looking out the window of her home in Tallevast, Florida, a community of eighty-seven households located thirty-eight miles south of Tampa. Within minutes, Ward was out her front door and across the lawn asking the crew chief what he was doing. What she learned was that Lockheed Martin, the most recent owner of the high-tech weapons plant located just down the street, had hired the drilling crew to determine if toxic chemicals from their facility had seeped into the groundwater and spread beneath the homes of Ward and her three hundred neighbors.

When she heard there might be toxic chemicals in the groundwater, Ward felt her world shift beneath her feet. She and her neighbors depended on the shallow wells that had provided them with water for generations. The possibility that this water might be contaminated was truly frightening.

"I'm angry," says Ward, who with Wanda Washington leads a small community organization called Family Oriented Community United and Strong (FOCUS). "I made baby formula and cooked for my family with that water for years while people at Lockheed Martin and at the county regulatory agencies knew how harmful it was," she charges. Ward, who has two children who have had "bouts with cancer," says she is also upset because she had to learn about the contamination herself.

## Former Turpentine Camp

Before the drillers were spotted on Ward's lawn, life in Tallevast, an African American community, had been relatively quiet. Like many of the other residents in town, Laura Ward's husband, Clifford ("Billy") Ward, the town dentist, traces his family's history on the land back to the 1890s when it began as a "turp camp," where freed slaves found jobs teasing sap out of the long-leaf "slash" pines and boiling it up into turpentine. As a youth, Ward and his father worked as migrant laborers following the harvest from the vegetable and fruit crops of Florida up to the apple orchards of New York State. Other Tallevast residents stayed closer to home, working in the orange groves, on dairy farms, and for Ringling Brothers Circus, which is headquartered nearby.

It wasn't until 1948 that Visioneering, a small machine shop, opened its factory across the street from the Ward home just down the street from the town's one store and post office. Initially residents appreciated the relatively well-paying maintenance, janitorial, and machinist jobs the new plant provided. Making the jobs even more attractive was the fact that residents could walk to work and then return home for a hot lunch at noon. For the most part, no one in town paid much attention to the factory next door where "metals were milled, lathed, and drilled into various components. Chemicals used and wastes generated at the facility included oils, petroleum based fuels, solvents, acids, and metals."[1] A decade later, the machine shop was renamed American Beryllium Company (ABC), and it began to handle larger quantities of toxic materials to make weapons.

By 1961 the Loral Metal Company completed its purchase of ABC, which it had begun to acquire it 1957, and renamed it Loral American Beryllium (LAB). At that time the Cold War was heating up as the arms race with the Soviet Union shifted into high gear. Recruitment of highly skilled machinists, hired to perform the ultraprecision metalwork, accelerated at LAB. The skills of the new workers were required to fabricate parts for nuclear weapons, atomic reactors, and space program projects under contract with the U.S. Department of Defense and the U.S. Department of Energy. Then in 1966, Loral American Beryllium was purchased by Lockheed Martin.[2] If they had it to do over, Lockheed Martin officials would probably decide against this acquisition because the purchase came with a major environmental liability, which they likely will be paying for over the next several decades.

## Residents Not Told of Contamination

Lockheed Martin employees first reported contamination problems when they discovered that a sump pump in Building #5 had broken and spilled large quantities of industrial solvents and cancer-causing chemicals into the soil and groundwater. Among the chemicals of concern found in the groundwater that exceeded Florida's Department of Environmental Protection (DEP) concentration guidelines were tetrachloroethene, 1,1-dichloroethene, beryllium, and chromium. Soil samples also contained excessive levels of volatile organic compounds (VOCs), total petroleum hydrocarbons (TPHs), as well as other compounds and metals.[3]

Lockheed Martin officials dutifully reported the contamination to local environmental officials at the Manatee County Environmental Management, but they did not feel obligated to inform local residents of the spillage. This lack of candor with local residents is now part of a lawsuit. Residents did not learn they were living atop a spreading plume of toxic chemicals until three years later.

From July 2000 until October 2003, Lockheed Martin officials engaged in a quiet, voluntary cleanup of some of their on-site contamination. They hired Tetra Tech, a California-based company, to remove 538 tons of tainted soils. In the course of their environmental sampling, Tetra Tech employees found a number of highly toxic chemicals, including trichloroethene, tetrachloroethene, dichloroethene, dichloroethane, and vinyl chloride.[4] They subsequently informed state environmental officials that contaminants from the five-acre company site, including trichloroethylene (TCE) and a number of other solvents, were migrating off site in the groundwater.

## Inadequate Regulatory Response

Many residents of Tallevast suspect that their race partially accounts for the failure of regulators to adequately protect them. "This would have been handled differently in a white community," says Wanda Washington about the contamination found in the drinking water in her home town. "I think it is because of skin color. . . . The government needs to be schooled that it is not all right to bring this type of facility into residential communities," she continues, referring to the Lockheed Martin weapons plant across the street that is the source of the pollution.

Washington, forty-eight years old, describes herself as a quiet woman, a mother of three, and a database administrator with a degree in psychology. But since some well water in Tallevast has been found to contain 250 to 500 ppb of trichloroethylene, a known carcinogen for which the regulatory standard is 3 ppb, Washington has begun to speak out: "I'm on the frontline because this contamination affected my family. I have no choice. This story needs to be told."

"I hope God protects us and builds a fence around us," says Washington. But she knows that the shallow wells that residents used for decades have already brought poisons into their homes. "Even if we can't prove it scientifically, that these chemicals are causing cancer in our community, we all know it," says Washington who is outraged that county officials knew about the contamination for three years before residents found out about it on their own. "Who is looking out for us? Everybody knew except us, and we are living with these poisons. I can't believe that anyone can be so cruel."

## Health Impact

Washington is convinced that the pollution has already devastated the health of her family. Her seventy-year-old mother, Lillie Flemming, has breast cancer that is being treated with chemotherapy. She also has diabetes, skin growths, and a cough that doctors cannot treat effectively. "I'm upset about what is happening. In fact, I'm angry as hell. I made formula out of water and fed it to my children," says Flemming. "I had one child who died at seven months, one who was retarded, and two who survived."

Washington's sister, Robin Darville, thirty-eight years old, also suffers from a number of ailments that could have been caused by the pollution. She had a stroke that left her with memory loss so severe that at first she didn't recognize her mother or husband. "I had to learn to walk all over again," says Darville, who also gave birth to an underweight infant. Since her stroke, she has experienced migraines, nonepileptic seizures, and difficulty grasping objects in her right hand. Darville's doctors are puzzled about why she has so many health problems at such a young age. "I think it is because of the contamination," Darville suggests. "This is personal. Lockheed Martin should pay," she adds.

### Amateur Environmental Investigators

With regulators providing residents with minimal information about the spill and its potential health impact, Ward and Washington felt the need to become amateur environmental investigators and began traveling to county and state regulatory offices to see what they could learn about the contamination. What they discovered was that a county official had been dispatched to see if any residents in the community were using their own wells, but the official failed to get out of her car to check because she was afraid of dogs, county records indicated.

State officials later conceded that the "notification provisions of our rules were not adequate" and that residents should not have had to wait three years to hear about the contamination under their homes. The Florida legislature later passed what is known as the Tallevast rule, which requires regulatory officials to promptly inform affected residents when a contamination problem is discovered. This fix, however, came too late to help Tallevast residents.

Once news of the contamination appeared in the media, county officials arrived in Tallevast the same evening and handed out five-gallon plastic bottles of water while warning residents not to drink tap water if it came from a well. Subsequently all wells in town were capped, and above-ground blue plastic pipes were installed and hooked up to county water lines. (These "temporary" blue plastic pipes remained in place four years later.) The increase in pressure from the county water hook up caused numerous leaks in their pipes faucets and hot water heaters in Tallevast homes, and the telephones of Laura Ward and Wanda Washington rang for days with requests for help with plumbing problems.

At a hastily convened town meeting at a church in Tallevast, Lockheed Martin officials assured residents that the danger from the chemicals that had invaded their drinking water was minimal—if it existed at all. But Ward, Washington, and other residents were not convinced by these reassurances and decided to do their own community health survey. For help they turned to a neighbor, Helen Worthington, a retired nurse who was well known and trusted in the community.

**Health Survey**

Helen E. Beyers Worthington married into the Beyers family, which traces its lineage back to the early days when the turp camp was still in operation in Tallevast. In those days, workers would toss potatoes into the cauldrons of boiling sap where they would cook instantly and bob to the surface covered in a glistening coat that shattered like glass when stuck with a fork. "We come from hard-working people who sent their children to college," says Worthington, who graduated from Texas A&M with a degree in nursing. A descendant of Thomas Jefferson and Sally Hemmings, Worthington joined the Air Force, married an Air Force man, and worked for forty-three years as a nurse at bases in Florida, Arizona, New Hampshire, London, and Guam, among other postings.

When approached about conducting a health survey Worthington was skeptical that she would find any problem. But after a couple of hours visiting with different families, she changed her mind. "There is something terribly wrong here," she told Ward. Sitting at a long table in the FOCUS offices across the street from Lockheed Martin, Worthington chronicled the health problems in her community, referring to her notes written in a careful, spidery script on a yellow legal pad. The heading at the top read: "Tallevast Florida: Tracking Household Illness."

Worthington was astonished by the number of residents with cancer in her small community. She recalls visiting one family where four of seven brothers died of cancer of the throat and other sites. Next door were three men in the family with cancer of the liver, and their sons were also having liver problems. In another family, Worthington found eight of ten children had died young of brain, lung, and uterine cancer and leukemia. "This was more than a coincidence," she opines.

On February 19, 2005, Worthington did a count of how many residents in Tallevast were living with cancer. She had to count herself as a cervical cancer survivor. Out of eighty-seven households, Worthington found fifteen had cancer and three of those have since died. This tally did not include those listed above who had already died of cancer or those who probably have cancer but prefer not to admit it. "In our community people do not like to admit that they have cancer. People here are very proud and do not like to talk about their problems," she explains. As an example, Worthington points to a young man in town with lymphoma who is living at home with his grandfather whom he has chosen not to tell about his health problem because he does not want to worry him.

Although it was not part of her survey, Worthington also noted a suspiciously high incidence of miscarriages, sterility, low birth rates, neurological disorders, and retardation. Other residents have health problems that will never show up in Worthington's study. Among them are Yvonne (Peggy) Ward's forty-two-year-old son who has to sleep with an oxygen mask strapped to his face, and Theresa (Pat) Robinson who has a similar problem with a daughter with "breathing problems."

"Almost every house in town has people with health problems, and it makes me angry that no one from the county or state was paying attention to them," she continues. Worthington thinks that the concentration of cancer she found in Tallevast deserves a more formal health study by county or state health officials. While Lockheed Martin and regulatory officials assure residents there are no health problems resulting from the contamination, Worthington does not believe them. "By the time the cleanup is finished, we all will be dead," she says. "People here are frightened, but they don't know what to do. I don't want to move because I own my home and I worked for it all my life. If I had known about the contamination earlier, I would have moved. It isn't safe to stay here because there are too many unknowns."

Worthington's neighbor, Fred Bryant agrees. "If I were younger I'd move," says Bryant, who has lived in Tallevast all seventy-eight years of his life and worked as a butler. Bryant does not want to move because he loves the close ties he has with other residents. "This is a place where when you cry, someone cries with you," he explains. So it is hard to leave. But Bryant does not want his grandchildren growing up on top of the contamination, which he thinks is causing illness. "You wonder who it will hit next," he says.

**Expanding Plume**

Not only did Lockheed Martin fail to inform residents promptly about the contamination, the company's declarations about having "found the edge of the plume" and delineated the extent of the contamination can best be described as serially optimistic. With each new phase of testing, the size of the contaminated plume of groundwater continued to expand like a drop of ink on a wet paper towel.

News about the size of the area contaminated by the Lockheed Martin facility unfolded slowly. In July 2003, state officials at Florida's DEP approved Lockheed Martin's Contamination Assessment Report, which indicated that most of the contamination was confined to their five-acre site with a small plume extending northeast of the facility. This assessment later proved to be inaccurate.

A time line of events in Tallevast was pieced together by *Bradenton Herald* reporter Donna Wright and later by Wilma Subra, a chemist who advises community residents on technical aspects of contamination problems. According to this chronology, Lockheed officials first described the contamination as confined to the company's property. However, the discovery of toxic chemicals in private drinking water and irrigation wells beyond the boundaries of the plant made it clear that the problem was larger in scale. By 2003 Lockheed Martin officials informed state regulators that the plume of toxics had crossed over into the residential community and worked out a cleanup plan. By May 2003, a Tetra Tech employee's report to Florida's DEP noted that the toxic plume had spread to a twelve-acre area off site.[5] This was confirmed in April 2004 when a sampling of water from seventeen wells by state and local regulatory officials found five wells outside the established plume of contamination with elevated solvent levels. The story about the spreading plume of toxics into the adjacent residential community broke in the *Bradenton Herald* on May 7, 2004. By the end of the month, state DEP and Department of Health officials discovered that the contamination was worse than they first thought.

### Elevated Exposure

In a report issued in July 2004, DEP officials reported that previous analysis of groundwater samples "indicated the presence of chlorinated solvents exceeding Florida Primary Drinking Water Standards (FPDWS)." Five irrigation wells and five supply wells located outside the plant property were found to be contaminated with TCE levels that exceeded state standards, and arsenic was also detected in a soil sample. State officials concluded that there was a "much larger chlorinated solvent ground water plume, with significantly higher concentrations of chlorinated solvents" than had been delineated by Tetra Tech.[6] State officials also found evidence that soil samples in residential areas exceeded the state's soil cleanup target levels for arsenic, barium, lead, benzo(a)pyrene, benzo(a)-fluoranthene and total recoverable petrochemical hydrocarbons.[7] One soil sample had 1,114 milligrams per kilogram of lead, whereas the state standard is 400 milligrams per kilogram. Soil samples were also taken from residential properties where dirt taken from the American Beryllium Company property had been used as fill.[8]

By June 10, 2004, the results of sampling of well water in the residents around the plant showed nine of twenty-four wells with traces of TCE. This sampling, paid for by local residents, revealed that two of the homes had 116 times the level of TCE considered safe, Wilma Subra reports. On July 23, 2004, testing samples from the plant site revealed "a reading of the solvent trichloroethylene at ten thousand times the drinking water standard," continues Subra, who believes that the toxic plume may have originated from multiple sites of contamination on company property. A few weeks later, on August 21, 2004, Lockheed Martin officials released a report showing that TCE "solvent levels in water beneath the plant at nearly twelve thousand times the state standard, and in nearby wells at up to five hundred times the code," Subra adds.

A year later, in June 2005, after further test wells had been drilled, the company recalculated the contamination as covering 131 acres. Since then the estimate has risen to some 200 acres, and still there are questions about contaminated wells beyond this area. Recently the Florida DEP ordered that the cattle herd of Heidi Boothe, on a farm near Tallevast residents, be tested after independent tests found the degreaser 1,4-dioxane, in her well water.[9]

Using funds made available to them by Lockheed, FOCUS hired an independent technical advisor, Tim Varney, who works with the Environ International Corporation in Tampa, as well as Michael A. Graves, a geologist who works for Environmental Sciences and Technologies. After sampling thirty-five drinking water and irrigation wells, Graves says his testing reveals a "deep diving plume" that has not been adequately delineated and is moving faster than previously thought. The plume reaches almost to U.S. Route 301 and may have reached the Floridian aquifer system, which provides drinking water to the majority of people in the state, he adds.

## Minimizing the Problem

While conceding that the chemicals had spread into the groundwater in the surrounding area, Lockheed Martin officials denied that there was any threat to the health of local residents. "Let me reaffirm that our company is committed to doing the right thing for the residents and has acted responsibly to uphold that commitment," writes Kenneth H. Measley, Lockheed's vice president of energy, environmental, and safety.

But Wilma Subra, a recipient of the prestigious McArthur "genius" prize for her work with contaminated communities, is less sanguine about the health threat posed by the Lockheed Martin plume. Residents may have been exposed to contaminants by drinking well water and by chemicals in the soil vaporizing and infiltrating into their homes, she explains. "The contamination is under the residential area, and it is at a very shallow depth," she points out. "These chemicals are very toxic. These people should not be living over the groundwater plume. To have residents living on top of this plume is putting them at risk. . . . You have to get them out of there," she continues.[10] "This is not a small plume. It is a dangerous plume. It is very deep in some places and very shallow in other areas, and it is under a residential neighborhood. The groundwater below residential areas is less than five feet below the surface."[11]

While some residents were initially hesitant to talk about their health concerns for fear that news of the toxics problem in town would bring down property values, over time they began to speak out. By April 2006, four hundred residents participated in an event organized by local activists at which a health survey was launched. State officials had declined to pay for a health study, so residents organized their own and have requested that Lockheed Martin officials pay for it. Residents "hope the survey will support their contention that pollution from the former weapons plant is responsible for an unusually high rate of cancer, miscarriage, and other ailments in the community of 80 homes," a local paper reports.[12] "How can you ingest this TCE without having consequences?" asks Lewis Pryor, a local resident who suffers from diabetes and has no history of the disease in his family.

## Routes of Exposure

There are various possible routes by which residents of Tallevast might be or might have been exposed to toxic chemicals that leaked off the Lockheed Martin site. Contaminated water was pumped up from the shallow aquifer and came into the homes of Tallevast residents and out of their faucets. They drank the water, cooked with it, bathed in it; their children played in the water from garden hoses; and they used it to wash their cars and water their lawns.

Another possible route of exposure is through inhalation of toxic gases. Lockheed Martin officials agree that VOCs in groundwater are a potential concern, but they do not consider inhalation a significant exposure pathway because air concentrations are presumed to be low. "Such a presumption is not acceptable," observes Subra, who is asking that more vapor intrusion studies be done in homes when water is in use. Both indoor and outdoor air samples should be taken when faucets are on in home sinks and showers and when outside irrigation systems are active.

Residents were also exposed to toxic chemicals in dirt that came from company grounds spread on the yards of some homes as fill when the plant was owned by the American Beryllium Company. Beverley Bradley, a postal employee for twenty-four years who has lived across the street from the plant her entire fifty-two years, remembers when three truckloads of dirt from the facility were spread in her backyard. An avid gardener who likes to grow flowers and tend her orange and banana trees, Bradley has dark skin lesions on her hands, arms, and feet. "The problem spread and has never gone away," she explains. "I think the contamination may have caused it," she adds. Bradley had four miscarriages, one child who was stillborn, one who survived for a few hours, and only one who survived. "As kids we played in the drainage ditches that came from the plant," Bradley recalls. Company officials came to town offering jobs, "and it was something good. Only later did we find out it was not so good. Many of us now feel betrayed and used. I'm angry. My family worked hard to own property to give me, and I want to give it to my son. But now, because of the contamination, we can't even get a loan to fix up the house or build a new one. We are stuck in a bad situation with no solution. An injustice has been done, and someone should pay," she adds.

## "We Did the Messy Work"

Residents and workers were also exposed to beryllium dust, which causes berylliosis, a disease that results from exposure to the heavy metal. Berylliosis can cause respiratory and central nervous system problems as well as attacking other organs. According to the accounts of Tallevast residents who worked at the plant, most of them were hired for janitorial and maintenance jobs, while only a few were employed as machinists. "We did all the messy work," says Bruce Bryant, fifty-five years old, who worked at the plant for six years. The dirty work involved milling large chunks of beryllium used in nuclear weapons and cleaning the beryllium dust out of the vents and the plant's attic. "None of us knew what beryllium could do to you," explains Bryant who is recovering from surgery for a cancer that has spread to the lymph nodes and bladder. "The cancer could be related to my work at the plant," he conjectures.

Sitting with Bryant on folding chairs at the Mt. Tabor Missionary Baptist Church were six other former employees of the American Beryllium Company: Anthony Smith, Walter Bryant, Norris Bryant, Errol Darville, Clarence Byers, and Morris Robinson. Two of them have cancer, and one needs oxygen to help him breathe at night. A number of their coworkers have already died of cancer, among them Ernest Smith and Anthony Smith's brother, who died of throat cancer at age twenty-nine. "I didn't feel great about working there but I needed the work," says seventy-eight-year-old Robert Smith, who worked at ABC for twenty-nine years and now has central nervous system and balance problems. None of the former workers gathered at the church is receiving compensation for health problems related to their work at the plant. "It seems like you have to be dying to get any money," Bryant observes.

An environmental assessment of the Tallevast plant, conducted by beryllium experts in 1997, described dust residue at the plant as "one of the worst they had ever seen," Subra reports. County blood tests of 241 residents found 7 testing positive for beryllium sensitivity. These findings indicate that Tallevast residents and their families were exposed to beryllium dust, say Laurence Fuortes, an expert in the field from the University of Iowa. A later accounting of local beryllium exposure reveals that ten local workers tested positive, as well as five local residents who did not work at the plant.

Some of the older surviving employees think that working at the plant was lethal for many. "Too many people I worked with are dead. A lot of young ones had nervous system disease and incurable illnesses," says seventy-eight-year-old Clarence Byers, who worked for nine years at the plant as a machinist. One of the men who worked at the plant, who just turned fifty, used to be a great catcher on a local baseball team and was an excellent athlete, Beyers recalls. "Now he is in a wheelchair because he can't walk and he can hardly talk . . . his words are all slurred," he continues. All the janitors Beyers worked with are now dead, he states. "It's lucky that every damn one of us is not sick," he comments.

Machinists who made a mistake working on a hunk of beryllium would sometimes ask a janitor to take the scrap piece and throw it into the plant pond to hide their mistake, Bryant recalls. Apparently unaware of this practice, plant officials held an annual public Fishathon during which they stocked the pond with trout so that workers and their families could catch fish to take home to eat.

In addition to eating fish likely contaminated with beryllium and breathing in the dust that blew off the factory roof, residents were also exposed to beryllium through the dust brought home on the clothes of employees. One of those likely affected in this way is Beatrice Ziegler, seventy-one years old, whose husband, Charlie Ziegler, worked at the plant. "He emptied the beryllium dust for twenty-nine years and came home coughing, she recalls. "Now Charlie can't breathe good," Ziegler says of her husband who is undergoing surgery for berylliosis at a hospital in Oak Ridge, Tennessee. Her husband is not the only one in her home with berylliosis: both Beatrice Ziegler and her brother, who also lived in the house, have been diagnosed with the same problem. "I went to my doctor, and he said: 'Beatrice, you are full up with berylliosis and you got it from your husband.'" Ziegler says her main symptom is that she is "short of breath" and "some days I can barely walk." Her husband has an oxygen tank by his chair and his bed to help him breathe. Sleeping is hard because her husband suffers at night. "I wait to sleep because of him, and if I am hurting I don't tell him," she adds. Ziegler says she and her brother are not receiving any help with their medical treatment from the plant because they never worked there.

## Incomplete Cleanup

Plans are being made to clean up the site, a project that is expected to last over twenty years. Lockheed Martin officials hired Blasland, Bouck & Lee, an environmental cleanup firm, to oversee its remediation plan, which is pending approval by state regulators. As a first step, the company plans to install 60-gallon-a-minute pump-and-treat equipment that will use titanium dioxide and intense ultraviolet light to treat polluted water, explains Tina Armstrong, Lockheed's senior project manager for the Tallevast cleanup. The treatment system is said to be effective at destroying TCE and other industrial solvents. The groundwater will be extracted by ten wells, pumped into a 21,000-gallon tank, treated to state standards, and released into the sewer system.[13] Blasland engineers will also search for globs of nondissolved contaminates—known technically as NAPLs (nonaqueous phase liquids)—that may hinder the effectiveness of the pump-and-treat system. There are also plans to drill more off-site test wells to determine how far and how deep the contamination of industrial solvents has spread, notes Gail Rymer, Lockheed spokesperson.[14]

But many residents are unsatisfied with the cleanup plan and want to be relocated. On January 21, 2005, Tallevast residents demanded that county officials relocate them and buy out their homes because of the contamination. Six months later, commissioners demanded that Lockheed Martin pay to relocate residents to safeguard them from health risks posed by the underground toxics plume. But since county lawyers began to fear that the county might be sued, the commissioners have not been heard from. "It is like they took the silent pill," says Brenda Pinkney, a forty-eight-year-old Tallevast resident who works as a counselor at a community college. After she informed her doctor about the chemical contaminants found in her groundwater, Pinkney's doctor advised her to get out of town. "Every day it gets scarier," says Pinkney, who lost her hair recently and now wears a wig. "I am afraid Lockheed Martin will admit its mistake too late. They are just going to let us die," she adds.

Lockheed Martin officials say there is no reason to uproot the community because residents are not at risk. After having delineated the 131-acre plume with 137 monitoring wells and 468 soil samples, company officials argue that cleanup efforts should be confined to the plant property and that soil and water samples of the plume off site show that it poses no threat to health and does not require remedial measures. Despite these assurances, state Representative Bill Galvano told county commissioners that federal, state, and county governments should come up with $20 million to move the 238 Tallevast residents living near the plant to new homes.

Lawsuits against Lockheed Martin have been consolidated under the direction of Motley Rice, a law firm in Mount Pleasant, South Carolina.[15] In all, 254 Tallevast residents have joined the suit. The lawyers for the Tallevast residents attempted to get the suit moved to the Twelfth Judicial Circuit in Manatee County; while Lockheed Martin attorneys want to keep the case in Tampa federal court. Company lawyers maintain that Lockheed Martin has "no responsibility for residents' alleged property damage or illness because work performed at the plant was done for the federal government," Subra writes.

**Power Imbalance**

In Tallevast, as in other fenceline communities around the nation, huge power imbalances exist in communities located adjacent to high-emission industries. Grassroots groups such as FOCUS are at a disadvantage when they confront wealthy corporations like Lockheed Martin over their pollution emissions. It is no coincidence that dirty industries do not set up shop in affluent areas: residents with both money and political connections would not put up with the pollution that poorer communities endure.

This power imbalance is on open display in Tallevast, where residents are determined to force Lockheed Martin to clean up the contaminated groundwater, pay for the damage to the health of residents, and either compensate them for the damage done to their property values or pay to relocate them. Achieving this outcome will be an uphill struggle: successfully suing large corporations for health damages caused by their emissions is the exception rather than the rule. The legal deck is stacked against residents, who bear the burden of proving that chemicals the company released caused illness in their community.

The power imbalance in Tallevast has caused a number of reactions. Some residents are demanding compensation and relocation, while others are pessimistic about their chances to be compensated for their losses. There is also a certain amount of paranoia. I spoke with a number of Tallevast residents who worry that speaking out against Lockheed Martin, which has extensive involvement in surveillance programs, makes it likely that they are being tape-recorded whenever they speak, spied on, and followed.

"It's gotten to the point where I can't take a shower without feeling that they are spying on me," says Wanda Washington. Washington describes herself as having grown up as a person who always saw the best in others. Before she encountered the contamination problems, "you had to prove to me that you were evil," she says. But that has changed since she has been dealing with corporate and regulatory officials. "Now I keep second-guessing myself," she says about meeting people. "Maybe I was just naive in the past." Washington is not alone. A number of residents I spoke with said they figured that the conversation that we were having at the FOCUS offices was bugged.

Such paranoia (or realism, depending on how you look at it) is understandable considering the size and specialized expertise of a corporate giant. In 2004, Lockheed Martin brought in $35 billion in revenues, $17.5 billion of it from the U.S. Department of Defense. This is the company that manages the information technology system for the Pentagon and provides intelligence gathering and fingerprint identification technology for the Department of Homeland Security, the Federal Bureau of Investigation, and the Transportation Security Administration, among other agencies. "Lockheed Martin will be involved in gathering information on the identities of millions of people in the United States as well as millions of tourists entering the country" through a passenger profiling system, a Polaris Institute report states.[16] The company also builds ballistic missile systems, as well as the Tomahawk, Trident, and Hellfire missiles, and antiballistic missile systems; maintains the National Aeronautics and Space Administration space shuttle and the president of the U.S. helicopter fleet; produces systems for the F-35 Joint Strike Fighter and the F-16 multirole fighter jet; and is being paid to develop a new spy plane.

To land these contracts and help the company out when it runs into trouble, Lockheed Martin spent a staggering $55.3 million on lobbying between 1998 and 2004 and hired 108 lobby firms to work on 512 issues before fifty-nine federal agencies.[17] The company donated over $7 million in three election cycles and $915,929 to congressional candidates in 2004. It made donations to fifty-three of sixty-two members of the House Armed Services Committee, fifty-one of sixty-six members of the Congressional Committee on Appropriations, and twenty-seven of forty-nine members of the House Select Committee on Homeland Security.[18]

## Lockheed Martin's Environmental Record

Lockheed Martin's environmental record has also come under fire. Richard Girard at the Polaris Institute, one of Lockheed Martin's most persistent critics, writes that the company's "production facilities, past and present, have also inflicted damage on the land and people who lived and worked near these plants. Much of the pollution from the production process occurred during the Cold War era when weapons manufacturing reached a peak rate and environmental laws in the United States were less stringent. . . . Even though the cold war ended over a decade ago, Lockheed Martin earmarked $420 million in 2004 toward cleaning up the mess it made during the last 50 years."[19]

Given this record, Tallevast residents are not the first (and will likely not be the last) community located near Lockheed Martin facilities to sue the company for environmental contamination. Among the cases that most closely resemble the contaminated plume in Tallevast are ones in Burbank and Redlands, California. In Burbank residents won their suit against Lockheed Martin for polluting local groundwater due to improper disposal of industrial solvents and toxic chemicals at its Skunk Works plant. In all the company paid $66.25 million to residents who claimed that the contamination had caused "various illnesses including breast cancer, leukemia, and non-Hodgkins lymphoma." The company was also ordered by the court to pay most of the $60 million cleanup costs.[20] In Redlands, the California Regional Water Quality Control Board ordered Lockheed Martin to clean up soil and groundwater pollution originating at its facility there at an estimated cost of $180 million. Some eight hundred residents sued the company in 1996, arguing that it had fouled their drinking water with trichloroethylene and perchloroethylene; more residents have since joined the suit. The California Supreme Court ruled in 2005 that each resident must individually "prove they are entitled to medical monitoring."[21]

## Quiet Strength

What is striking about the behavior of Tallevast residents in standing up to this giant weapons and surveillance company is how temperate and patient they have been after discovering that they had been drinking contaminated water for decades. "We didn't grow up demonstrating. We were taught you could get more with honey than with vinegar," one resident explained.

While residents have begun to speak out about the health problems they think were caused by the contamination, there have been no marches to the county offices or civil disobedient actions. What they have done is organize regular meetings to discuss the latest news about the investigation into the environmental hazard, hold candle-light vigils, and put up signs along the road at either end of town saying "Welcome to Toxic Tallevast" and "Leaving Tallevast, Decontaminate."

"This is a God-fearing, church-going community. We are strong in our beliefs and we know we will be all right but we need to expose what happened here," observes Reverend Willie Smith, the associate minister of Mt. Tabor Missionary Baptist Church, which was founded in 1907 and has a congregation of 150. "There are supposed to be checks and balances in our system of government to protect us from these kinds of problems . . . but nobody checked," notes Smith wryly, adding that in his view, it is mostly communities of color that are the focal points for the worst contamination problems.

Echoing Smith's view about the racial component of the problem is Reverend Charles S. McKenzie Jr., who describes himself as a spiritual advisor to FOCUS, helping the group with strategic decisions as well as providing a national perspective. The type of contamination story playing out in Tallevast "is often embedded in communities of color," notes McKenzie, a state volunteer coordinator for Reverend Jessie Jackson's nonprofit PUSH. "These communities become dumping grounds and don't have the political clout to stop it," he continues.

McKenzie credits Laura Ward and Wanda Washington for uncovering the contamination in their community and doing a good job of bringing it to the attention of local reporters while at the same time noting that the struggle for relocation has yet to grab the attention of the national media. Ward and Washington have done a good job of "channeling the anger in town in creative ways" as well as being "good models and mood-setters for the rest of the community," he adds.

"In the black church we are overcomers no matter how large the Leviathan. While Tallevast residents face a David-and-Goliath contest, it is not size but justice that will ultimately prevail," McKenzie predicts, lapsing into a preacher's cadence. "The strength and ability of community residents has to do with the rightness of their cause. Many battles are won not by might but by right," he adds. While McKenzie believes that a "providential hand" will help Tallevast residents prevail against one of the biggest weapons firms in the world, he also hopes to raise the profile of the Tallevast struggle with Jessie Jackson and with members of the Congressional Black Caucus.

While preachers talk about "David-and-Goliath" contests and the "providential hand," Clarence (Billy) Ward, the town dentist, is looking for something more worldly. "I'd like to see Lockheed Martin stand up and admit that a wrong has been committed and that they are sincere about making it right," he says in the office building next to his dental surgery, which he and his wife have given over to FOCUS. "Our roots are here, and it is unfair that we are being driven from our homes because of the neglect of others. No one can make up for the blood, sweat, and tears that have fashioned the Ward home," he continues. Nevertheless, he thinks it only right that Lockheed Martin should foot the bill for the relocation of all residents who want to move, as well as "fair compensation" for the health problems and loss of real estate equity that resulted from the contamination spread from its facility.

Four years ago, Ward asked Lockheed Martin officials for a list of other communities where it has contamination problems so that he could visit them and see how other communities on the fenceline with their pollution problems have been treated. Corporate officials never provided him with such a list, he observes. "Lockheed Martin officials have been paternalistic toward us by telling us what we need," says Ward who notes that they want residents to go to company doctors rather than their own. "But we are not all asleep here," he says with a glint in his eye.

## Double-Edged Sword

There is a tragic irony in the fact that by-products of the weapons being forged in the facility next door to Tallevast residents will most likely one day be proven to have caused illness and death among the people they were designed to protect. Residents of the neighborhoods surrounding Kelly Air Force Base in San Antonio, Texas, suffered the same plight as toxic chemicals dumped in open pits on the base entered the groundwater and seeped beneath the homes of thousands of neighboring residents. Their story is told in the next chapter.

Could the accidental contamination of these residential areas have been foreseen and averted? The use of the precautionary principle might have allowed responsible regulatory officials to have both foreseen and averted these outbreaks of environmentally induced disease because it would have suggested that permitting large volumes of toxic chemicals to be handled and used in an area immediately adjacent to a residential population was unwise. Either the facility or the residents should have been moved or, at the very least, special measures should have been taken to see that the disposal of toxic wastes was done in a manner that avoided residential exposures. Unfortunately the precautionary principle is rarely used in setting land use policy in and around our military bases and military-industrial facilities. As a result, the health of the people who reside near these facilities is often sacrificed as a kind of collateral damage that our defense system extracts.

# San Antonio, Texas: Contamination from Kelly Air Force Base Suspected of Causing Sickness and Death in Adjacent Latino Community

A purple wooden cross stands outside the white clapboard home of Guadalupe and Robert Alvarado Sr. For the past thirty-seven years the Alvarado family has lived across the Union Pacific railroad tracks from Kelly Air Force Base in San Antonio, Texas. The cross on their lawn, one of many in this working-class, largely Latino community, signifies that someone inside is either living with cancer or has died from it. Residents who erect these crosses suspect that the numerous cancers, kidney and liver problems, neurological diseases, and reproductive disorders found in their community were caused by pollution from the base.

Guadalupe Alvarado suffers from thyroid cancer and diabetes, ailments she thinks were caused by the huge toxic plume of chemical contaminants that leaked from the base over several decades. These toxins, some of which cause cancer, seeped into a shallow aquifer and spread for four square miles under the homes of an estimated twenty-two thousand to thirty thousand residents in the surrounding communities.

## Health Impact

"A lot of us witnessed loved ones dying. I hurt when good friends died, especially when they were people you grew up with. It hurt when they started falling sick with cancer and had to pay for chemo [therapy]. Some were young and never had a chance to live a real life. It hurts now when we have young ones coming down with cancer and asthma. We didn't ask for any of this. On my street fifteen people died of cancer. We put up purple crosses. Now we see more and more purple crosses for kids," says Guadalupe Alvarado. Her own daughter, Lisa, thirty-seven years old, has thyroid cancer.

Guadalupe's sixty-five-year-old husband has advanced liver and kidney disease, an aneurysm that left him legally blind, and a malfunctioning thyroid. Confronted with this long list of serious medical problems, his doctors asked if he had been exposed to high levels of radiation. Alvarado told them that as far as he knew, he had not been exposed to radiation but that he did live near a military base. He suspects that chemicals that spread from the base are the source of his afflictions and his suspicions are not unreasonable given that a number of the chemical contaminants from the base are known to cause diseases from which he and his family suffer. Trichloroethylene (TCE), for example, a chemical used in large quantities on the base to clean metal parts, is a probable carcinogen linked to liver and kidney cancer, as well as birth defects.

State health officials found elevated levels of liver cancer and birth defects in the area above the toxic plume but say they cannot prove the increased incidence of disease was caused by the polluted aquifer. Despite the inability of health officials to prove a causal link between the toxic releases and high disease rates in the affected area, there are several troubling signs. Among them is the fact that the liver cancer rate in communities near the base are about double the expected rate and has remained so over the past ten years, reports Melanie Williams, a cancer epidemiologist at the Texas Department of State Health Services. One health study reports five hundred cases of diagnosed liver cancer among residents who have lived near Kelly AFB since 1995.[1] Another federal study found elevated kidney cancer rates in two zip code areas adjacent to the base.[2] Birth defect rates were also found to be two to three times higher than expected. [3]

The many shallow wells in the neighborhoods adjacent to the base provided one of the routes of exposure through which residents came into contact with the contaminated water in the aquifer, which lies five to thirty feet below their homes. Over the past decade some seventy-five of these wells have been capped by the Air Force. "We know the people used the wells for drinking water," says George Rice, a hydrologist. They were also likely used for watering gardens, washing cars, "and the children used the hoses the way children use hoses," he adds.[4] In other words they drank from the hose, sprayed each other, danced under the sprinklers, and splashed in plastic pools filled with the contaminated water.

"I have struggled long and hard to support my family," says Alvarado, who over the years has worked selling insurance and transporting human remains for the University of Texas Medical Center before ending his career working for Delta Airlines. "I was fifty-seven when I got sick. They cut my work-life short. Guadalupe had to go back to work at a dry cleaners even though she is sick also."

Alvarado's family's deep roots in the San Antonio area go back to the 1800s. His father, Joe Leyva, was killed on June 6, 1945, when a German artillery shell hit the landing craft that was ferrying him and other members of the 300th Engineer Combat Division to the beach in Normandy. Taking his stepfather's name, Alvarado married Guadalupe in 1959 when he was eighteen and she was sixteen. At the time he earned twenty-five dollars a week. With limited income, Alvarado was always on the lookout for cheap ways to fix up and expand his home. He built a number of outbuildings in the back yard to accommodate his family, which grew to five children and eighteen grandchildren.

## Contaminated Topsoil

One low-budget opportunity to improve his home arose one day while he was fixing a flat tire at a local garage. A driver of a dump truck loaded with topsoil hauled out of Kelly AFB approached him and asked if he wanted to buy it. Alvarado was interested because his yard flooded during heavy rains and the price was right: fifteen dollars a load. Building up the yard with some topsoil would end the flooding and give his children a level space on which to play. So he paid for two loads and spread it over his lawn. At the time the transaction seemed innocuous enough, but Alvarado now sees the purchase as a tragic mistake because it exposed him and his family to toxic topsoil from the military base.

The health problems of Alvarado's family are not unusual in the neighborhoods surrounding Kelly AFB, one of seven military bases located in San Antonio. A quick walk around a few blocks in East Kelly reveals dozens of knee-high purple crosses planted in the lawns and next to driveways. Alvarado ticks off the toll of immediate neighbors with cancer: one died of stomach cancer, another of breast cancer, a young woman of twenty-five has stomach cancer, and Emma, who worked at the base, had a kidney transplant, Alvarado told a reporter.[5]

Elevated levels of disease are also occurring on other streets in the area. San Miguel, a sixty-year-old retired wrecker driver who lived near the base for twenty-seven years, was diagnosed with thyroid cancer in 2003. His neighbors are also sick and dying. "'The woman in that house has cancer,' he says. 'The one next to her has breast cancer, and another one over there has leukemia. . . . It's too many problems for one short block. It's not normal,'" he told a reporter for the *San Antonio Current*.[6] In addition to the cancer cluster are reports in the neighborhood of high levels of miscarriages and birth defects, central nervous system disorders, anemia, elevated asthma rates, and over 120 cases of amyotrophic lateral sclerosis (Lou Gehrig disease).

Officials first noticed something seriously wrong in the communities surrounding Kelly AFB 1989 when a group of construction workers fainted in 1989.[7] Rescue workers called to the site found that the workers had keeled over from fumes wafting up from toxic wastes they had unearthed while excavating along Quintana Road just outside the base's perimeter. Prior to the incident, Alvarado's neighbors had repeatedly complained about foul odors coming from the huge jet fuel tanks located on the base across the railroad tracks from where they lived. There were also complaints by residents whose fingernails turned black and hair fell out as a result of watering their lawn with water drawn from shallow wells. But no one listened to them until the utility workers collapsed.

**Large Volumes of Toxic Chemical Releases**

Where the plume originated is no mystery. Air Force officials concede that a wide variety of toxic chemicals used on the base made their way into the aquifer. One report claims that in a single year, as many as 282,000 tons of hazardous waste were generated at the base.[8] Whatever the exact figure, chemicals were disposed of in a primitive manner for decades on the 4,600 acre Kelly AFB—with its six hundred warehouses, machine shops, metal stripping shops, paint shops, and jet fuel storage tanks. The base, founded in 1916, was transformed into an Air Force base in 1940 and became the longest continually operated Air Force base until it closed on July 13, 2001. During the Korean and Vietnam wars, Kelly AFB was a major hub for maintenance work and storage, and at its peak, it employed some twenty-five thousand civilian workers while handling 50 percent of the Air Force's engine maintenance.

For decades, workers drained chemical wastes directly into the ground or dumped them in the creek at Kelly AFB, one of the largest sites contaminated with TCE in the nation. "One former worker admitted he was under orders annually to drain vats of chemicals into the ground during the Christmas holidays," report Anton Caputo and Jerry Needham in the *San Antonio Express News*.[9] There was also an open acid pit at the base where heavy metals were dumped. According to local accounts, a refrigerator thrown into the acid pit would quickly dissolve. Heavy rains periodically washed the wastes into the sewers and creeks overflowing into East Kelly, a neighborhood that lacked storm drains.

Yolanda Johnson, who has lived in North Kelly Gardens since 1965, remembers children playing along the fenceline with the base just across from a waste pit that periodically overflowed into the surrounding residential neighborhood when it rained. Two of her children developed "bowed bones" and had to sleep for years with their arms and legs in casts. Johnson thinks this might have been caused by chemical exposure and points to the fact that some of her grandchildren were born with kidney disease and one had a missing rib.[10]

Elevated Exposures

In 1983 officials at Kelly AFB released statements indicating that dangerous chemicals—including the carcinogens TCE, perchloroethylene, benzene, and chlorobenzene—had been dumped in an open pit between 1960 and 1973.[11] This revelation made it hardly surprising that a study conducted in 1998 recorded TCE levels as high as 49,000 ppb on the base. Subsequent surveys found 10 to 100 ppb TCE in the groundwater plume that runs from the base beneath adjacent residential areas. Under current regulations, TCE in drinking water is considered safe at 5 ppb, but the standard may soon be made stricter so that only 1 ppb is considered safe.[12]

In addition to TCE, dichloroethylene, tetrachloroethylene, polychlorinated biphenyls, vinyl chloride, benzene, and thalium were also used and dumped at Kelly AFB. Soils in the affected area are contaminated with jet fuel, radioactive wastes, volatile organic compounds (VOCs), nitric and sulfuric acid, beryllium, and heavy metals such as lead. Leon Creek, which flows for three and a half miles through the base, is heavily polluted, and many chemicals exceed allowable limits in the sediment and fish.[13]

## Getting Organized

In 1993 members of the North Kelly Gardens Committee became alarmed about the possibility that toxic releases from the nearby base were causing what appeared to be elevated rates of cancers, liver and kidney disease, and reproductive disorders in their community. Unsure what to do about it, a couple of residents contacted the Southwest Workers' Union (SWU), an eighteen-year-old union of public school janitors, maintenance workers, and gardeners. Contacting SWU had a number of advantages. For one thing, it provided Kelly residents with access to full-time, experienced organizers, whose help they would need if they were going to go toe-to-toe with officials from the Air Force base. Second, bringing SWU in to work on the contamination problem was a good fit because the union was well known and trusted by many local people who were already members or knew someone who was.

After being invited to meet with concerned residents, Genaro Lopez-Rendon, codirector of SWU, devoted fifteen years to working with Kelly residents on the contamination problem and during that period recruited some 350 residents to become SWU members. By 1994, a number of Kelly residents, including Robert Alvarado Sr., founded the Committee on Environmental Justice Action (CEJA), which would serve as the grassroots group of local residents directly affected by pollution from the base. With four hundred members enrolled, CEJA and SWU began organizing protest marches, conducting heath surveys, placing signs along the creek warning people not to fish, and educating over a thousand residents about the potential health effects of pollution from the base.[14]

## Capturing Evidence

A key breakthrough occurred when one Kelly resident, who joined SWU and lived near the base, agreed to have a well dug on his property and have the groundwater tested. Water from this test well turned out to be polluted with TCE, which causes liver and nervous system damage, and tertrachloroethylene, which causes liver and kidney damage. These chemicals, used to strip oil and paint from metal parts, degrade into 1,2-dichloroethene, a possible cancer-causing agent.

One of the early complaints of residents in East Kelly concerned odors coming from three jet fuel tanks, which held 3 million gallons of fuel. Despite assurances from Air Force officials that no contaminants were leaving the base, evidence of benzene pollution was detected and SWU and local residents pushed for the demolition of the three giant tanks. In 1998 they won their battle, and the tanks were demolished. "When the tanks came down, it gave people confidence that they had the power to make change," Rendon recalls.

Going from house to house, meeting residents around kitchen tables, Rendon, Jill Johnson, and a number of other SWU workers continued to organize. One campaign they mounted involved preventing the military from expanding its presence next to the East Kelly neighborhood where there were plans to station the RED HORSE division of engineers who specialize in earth moving. Employing a series of evening candlelight vigils, timed to take place after their members got off work, residents convinced military officials to station the additional troops elsewhere.

The major struggle, however, continued to focus on the groundwater contamination beneath the homes of local residents. Texas state officials acknowledged that "extensive environmental contamination at the base" and testing showed that the contamination had spread beneath twenty thousand homes. Air Force officials argued that "mother nature" would take care of the problem over time as a process of "natural attenuation" broke down the toxic contaminants into benign components.[15]

But local residents were unwilling to accept this as an adequate solution and demanded a more aggressive cleanup program. They also lobbied to have the area declared a federal Superfund site. To this end, in October 1999, CEJA, SWU, and the Texas Chapter of the Sierra Club filed a petition with Texas governor George W. Bush and the U.S. Environmental Protection Agency with the aim of winning a federal Superfund site designation for Kelly AFB.

Obtaining a Superfund site designation would be useful because residents in affected areas would be able to join the discussions with military and regulatory officials about how to solve the contamination problem, and they would be able to hire their own experts to advise them on cleanup proposals. Unfortunately, state officials opposed the application for a Superfund site designation. Blocked again, the SWU and CEJA filed "a civil rights complaint alleging discrimination against the Latino community near the base."[16]

## Help from Outside

In addition to the organizing help that SWU provided, residents needed assistance understanding the highly technical reports about the contamination. It is not unusual that communities faced with contamination problems are snowed under with a blizzard of technical reports written in scientific and bureaucratic jargon. Faced with this, some grassroots groups are unable to respond in an effective and timely manner to advocate for their interests. Fortunately, assistance with this task was provided by Wilma Subra, a chemist based in Louisiana who helps grassroots groups translate complex technical reports and regulations.

Prying loose data about the extent and severity of the contamination problem surrounding Kelly Air Force base was difficult. Even after some sampling of the toxic plume had been conducted, residents were not given the results of the testing and a Freedom of Information Act request for the information was denied. After trying, without success, to access documents from the Air Force, Subra used her contacts at the EPA to obtain the reports and then wrote up a series of handouts for residents that made clear what they faced. Through EPA documents, Subra learned that the groundwater beneath the base was heavily contaminated with numerous chemicals and heavy metals, which exceeded groundwater protection criteria. Among those in excess of regulatory standards were TCE, PCE, arsenic, benzene, chlorobenzene, 1,2- dichloroethene, nickel, and vinyl chloride.

Subra told residents that this heavy concentration of toxic chemicals was likely taking a toll on their health. As she explained it, the health problems experienced by the Alvarado family and other residents in Kelly "are consistent with those expected from exposure to the chemicals of concern and chemicals present in excess of criteria levels in groundwater under the community, surface waters that flow through the community, and the sediment and fish in the water bodies that flow from Kelly AFB through the community. The community should not be made to continue to live on top of this shallow, contaminated groundwater plume. The ongoing exposure is unacceptable," she asserts. "People living above this shallow groundwater plume are chronically exposed to these chemicals."

Since their shallow wells have been capped, residents are no longer exposed to toxic chemicals by using groundwater. However, they are exposed to these chemicals through inhalation, ingestion, and skin contact, Subra explains. Over the years, a large number of spills around the perimeter of the base combined with flooding left a toxic residue in the soil of nearby yards. Residents are also exposed to chemicals that came from the base through "inhalation from the off-gassing from contaminated soil and groundwater," she continues. One can actually smell this out-gassing process in the summer, she reports and concludes, "There is a need for a comprehensive and cumulative study addressing all routes of exposure."

In addition to holding a number of workshops in the neighborhoods surrounding Kelly AFB, at which she informed residents about that potential hazards they faced from chemical exposures, Subra also talked with elected officials about the problem and consulted with community members as a technical adviser on the Restoration Advisory Board. She also served on an Interagency Working Group jointly sponsored by the U.S. EPA and Texas Commission on Environmental Quality (TCEQ). The working group focused on environmental, human health, and economic development issues surrounding the cleanup of the contamination from the base. At many of these meetings SWU activists and residents in the "Toxic Triangle" criticized the slow pace of remediation on the base and the lack of medical help for affected residents.

### Incomplete Cleanup

To date the Air Force has spent $320.4 million on environmental investigation and cleanup at Kelly AFB, and that price tag could rise to $465 million by 2024, says Sonja Coderre, public affairs officer at the Air Force Real Property Agency.[17] Further tightening of TCE standards may increase the cost of the cleanup. The Air Force has also committed $10 million over five years for health tests at the neighborhood's Environmental Health and Wellness Center.[18] Despite these substantial outlays for remediation, only 475 of 687 potentially contaminated sites on the base have been cleaned up according to federal and state officials. Underscoring the untreated extent of TCE contamination at the base was the discovery of several drums of TCE, which were removed from an area under the fifteenth tee at the bases's golf course.[19]

To date, the Air Force has installed permeable reactive barriers that are buried underground and are designed to contain and filter out harmful chemicals.[20] Contaminated groundwater is also pumped to the surface, treated, and then released in a system known as pump-and-treat. Members of the SWU and resident activists argue that this approach is inadequate and demand a more aggressive on-site cleanup that would dig up, haul away, and eliminate the contaminated soil that is the ongoing source of off-site groundwater contamination.

In one of the most heavily contaminated sites on the base, the old metal plating shop, military officials have erected a cement wall around it to contain the toxins instead of removing them. "We're not for a containment plan. Those types of sites should be cleaned up immediately," argues Rendon.[21] Subra agrees: "The waste has been surrounded by a slurry wall and covered by an asphalt parking lot. This is not acceptable. The waste should be dug up, treated, and disposed of, not allowed to remain on site and continue to serve as a source of contamination for the groundwater resources."

Problems with the pump-and-treat system of mitigating groundwater contamination became apparent on October 5, 2006, when forty-five thousand gallons of water contaminated with chlorinated solvents spilled at Kelly AFB, contaminating areas outside and inside the water treatment plant. The accident began when an ultraviolet oxidation recovery machine, used to treat the water, shut down because of low water flow. Because of a computer error, groundwater from recovery wells continued to arrive at the treatment plant and overflowed its storage tank. Some thirty-six thousand gallons of groundwater contaminated with PCE, dichloroethene, and TCE were released, while the remaining nine thousand gallons remained within the treatment building. Long-term exposure to these chemicals can cause kidney and liver damage and cancer; persons exposed to high levels can faint; and short-term exposure can cause headaches, skin irritation, and drowsiness.

## Widespread TCE Spills at Military Bases

The twelve-square-mile plume of toxics from Kelly AFB that spread under adjacent residential areas is by no means unique in the country.[22] According to Air Force documents, nationwide fourteen hundred known military sites have TCE contamination problems.[23] At one of these, the Camp Lejeune Marine Corps base in North Carolina, drinking water was found to contain 1,400 ppb of TCE, according to the Agency for Toxic Substance and Disease Registry. The current EPA standard is 5 ppb. "A still incomplete study of 12,598 children born on the base from 1968 to 1985 found 102 cases of cancer and birth defects, including 22 cases of leukemia, twice the national average. No studies have been conducted of the adult men or women who drank the base water," Ralph Vartabedian reports in the *Los Angeles Times*. Elizabeth Dole, North Carolina's senator at the time, asked the Government Accounting Office to investigate whether the Marine Corps covered up the TCE problem at the base.[24]

The type of massive toxic spills on bases such as Kelly and Lejeune are the result of a longstanding pattern of irresponsible disposal of chemicals on military bases and a lack of adequate military regulations about how toxics should be handled in a way that protects the health of military personnel, civilian workers, and residents of adjacent communities. For too long, the military exempted itself from rules about the dumping of toxics, and as a result thousands of contaminated areas in and near military bases continue to cause health problems.

Rather than simply admit past mistakes and upgrade their environmental regulations, the military is involved in a rearguard action to minimize the toxics crisis on their grounds. Recognizing the scale of the TCE problem they face, the Department of Defense intervened in a regulatory debate with officials at the EPA over what level of TCE is safe in drinking water. After a review of the accumulating scientific and medical literature, EPA officials concluded that they should tighten regulations about exposure to TCE in drinking water from 5 ppb to 1 ppb. The Department of Defense promptly questioned the reliability of the EPA finding, and the National Research Council (NRC) was brought in to mediate the dispute. On July 27, 2007, the NRC issued a 379-page report siding with the EPA's risk assessment and recommending the more stringent safety standard. The NRC found the research on the potential of TCE to cause kidney cancer as particularly persuasive. The report also confirmed the existence of a 2001 EPA document linking TCE to kidney cancer, reproductive and developmental damage, impaired neurological function, and autoimmune disease.

**Collateral Damage**

Although this may sound like an arcane regulatory dispute between government agencies, the lack of enforcement of strict exposure standards can incur devastating health consequences. For example, Mary Lou Ornelias, Robert Alvarado Sr.'s cousin, described to a reporter how, during the eighteen years she worked on Kelly AFB, her job required her to dip cotton cloths into buckets of TCE with her bare hands and wipe grease from aircraft parts.[25] "I started working there young at Kelly," says Ornelias, who later had trouble breathing and was diagnosed with liver cancer at the age of fifty-two. "I liked my job. I didn't know it was going to affect me this way."[26] Following her diagnosis, she started throwing up blood in 2002 and died in September 2006. At her funeral, her son, Jacob Moran, noted that his mother had never joined lawsuits to seek compensation for damage to her health caused by exposure to TCE. "She just wanted it to be known that those chemicals were dangerous," he observed.[27]

The irony of the U.S. military's inadvertently killing and sickening thousands of American citizens by exposing them to toxic wastes, all in the name of protecting them, has not been lost on some of the self-described victims of military toxics. In an article entitled "Military Wastes in Our Drinking Water," Sunaura and Astra Taylor write: "In 2003, when the Defense Department sought (and later received) exemptions from America's main environmental laws, the irony dawned on us. The military is given the license to pollute air and water, dispose of used munitions, and endanger wildlife with impunity. The Defense Department is willing to poison the very citizens it is supposed to protect in the cause of national security." The interest of the Taylor family in the quandary posed by irresponsible disposal of military wastes came about because their daughter was born with congenital birth defects, which they are convinced was caused by drinking water laced with TCE that was dumped at the Tucson Airport by military contractors during and after the Korean War.

"Today the U.S. military generates over one-third of our nation's toxic wastes, which it disposes of very poorly. The military is one of the most widespread violators of environmental laws. People made ill by this toxic waste are, in effect, victims of war. But they are rarely acknowledged as such," Taylor and Taylor write. "It is an ugly truth that manufacturing weaponry to kill abroad also kills at home. The process involves toxic chemicals, metals, and radioactive materials. As a consequence the U.S. military produces more hazardous waste annually than the five largest international chemical companies combined."[28]

## Many Civilian Workers Sick with ALS

Cancer and birth defects are not the only medical problems likely caused by the irresponsible handling and disposal of toxic chemicals at military bases. Residents in the communities surrounding Kelly AFB are concerned that too many of them have been diagnosed with amyotrophic lateral sclerosis (ALS), a progressive neurodegenerative ailment, commonly known as Lou Gehrig's disease. There are some thirty thousand cases of ALS nationwide at any one time. Symptoms associated with the disease include muscle weakness in the arms and legs, slurred speech, difficulty swallowing or breathing, memory loss, headaches, joint pain, and chronic fatigue. Following diagnosis, those suffering from this disease live, on average, two to five years.

A total of 127 former civilian employees of Kelly AFB completed questionnaires indicating that they had been diagnosed with ALS.[29] That seemed to local residents to be an overly large number of neighbors to be afflicted with this relatively rare disease in such a small community. To determine whether there was a problem with ALS in the communities surrounding Kelly AFB, the U.S. Air Force undertook a mortality study. The researchers found that between 1981 and 2000, there had been thirteen deaths from ALS among the former civilian workers who lived near the base. This number was not found to be excessive. "There are not an increased number of deaths from ALS" among these civilian workers, the ALS Association reported.[30] The mortality study, however, excluded military service men and women who worked at the base who subsequently died of ALS. Nor did it include the 127 Kelly residents who reported having been diagnosed with the disease. They didn't count in this mortality study because they are still alive. "We are aware of the limitations of a mortality study, when it is conducted at a time when the majority of the population in question is still alive. With regards to ALS—most [Kelly civilian workers] have not reached the ages where the risk of ALS is maximal," Dr. Carmel Armon notes. He recommended a follow-up mortality study in five years.[31]

The existence of this ALS cluster next to Kelly AFB, if confirmed, will not be altogether surprising given that a number of studies found elevated levels of ALS among military personnel. A Harvard study reported in *Neurology* on January 11, 2005, revealed that male veterans had a 60 percent greater risk of developing ALS than men who had not served in the military. Two other studies report that Gulf War veterans are twice as likely to develop ALS compared with those not deployed to the Gulf. "Environmental factors may play a role, such as exposure to chemicals during military training," the ALS Association report speculates.[32]

### Military Contamination Conference

The contamination of the Latino communities surrounding Kelly AFB "is not an isolated incident . . . it is part of a larger pattern," asserts Lopez-Rendon, the SWU organizer. With six thousand U.S. military bases in the United States and territories and seven hundred bases abroad, "the U.S. military is the largest source of contamination in the world," claims his colleague Jill Johnson. To forge links among community groups dealing with these issues, SWU hosted a conference in July 2006: "The Converging Community Struggles for Health and Justice: Movement Building against Military Contamination." The conference was attended by antimilitary-toxics activists from all over the nation and the world.

At the culmination of the four-day gathering, participants, SWU organizers, local activists, and neighborhood residents marched through the 100 degree heat on the outskirts of southern San Antonio to the gates of the now-closed Kelly Air Force Base. The march was slow because a number of the participants, among them Robert Alvarado Sr., are seriously ill. Carrying banners demanding "Clean Up Kelly Toxics," the marchers were occasionally cheered on by cars and trucks that sounded their horns in sympathy with the demonstrators cause.

Once at the gates of the base, the demonstrators surrounded a jet fighter that remains behind as a relic of the base's military history. Speeches were given by activists from communities around the world that face similar problems. Change comes from working together with other communities, says Reuben Solis, an SWU organizer. "From this gathering we can see how big the problem is and it makes the heart heavy. . . . We have no time to waste. We have the power to make change," he concludes.

## Transparency

The next chapter describes a community where a public housing complex was built on and still operates atop contaminated soil from a public utility. This is the result of a tragic mistake that dates back to World War II. Despite the distant origins of the problem, there is evidence that the utility, county, and state officials knew there was a contamination problem long before residents were told. This lack of transparency is often discovered during the course of fenceline antipollution campaigns, and it was front and center in the struggle in Daly City, California, where residents in public housing adjacent to a Pacific Gas and Electric facility began to notice high levels of unusual illnesses among their family and neighbors.

The question of who knew what and when is an issue that is frequently heard in the parsing of Washington scandals and in lawsuits over cover-ups. In fenceline disputes, it is a crucial issue, because it gets at the question of whether residents have a right to know about contamination as soon as it comes to the attention of a facility operator or regulator. Often corporate and government officials are wary of releasing information about contamination that they feel might alarm residents who have no experience evaluating technical issues. This approach, however, misses the ethical dimension of the issue: that people should be given adequate information to make an informed decision when it comes to matters concerning their health.

# IV

## Contaminated Soil

# 9

## Daly City, California: Midway Village: Public Housing Built on Contaminated Soil

Lula Bishop moved into Midway Village Complex in Daly City, California, in 1978 and counted herself lucky at the time. The subsidized housing unit she occupied was located in a relatively desirable neighborhood one mile from San Francisco Bay, and Bishop was initially surprised at how attractive the housing was. True, her apartment was located across the fenceline from a Pacific Gas & Electric (PG&E) utility company maintenance yard, but the rest of the neighborhood was residential. There were also convenient facilities for her three children—Kevin, Kenneth, and Tonya—including a day care center, Head Start program, playground, park, and elementary school.

"It was a very pretty place, but it was surreal. Why would they put poor people here when they could have rented the units for two thousand dollars a month?" Bishop asked herself. "There was something wrong with the picture." Once she learned that her beautiful new home was built on toxic wastes, it suddenly made perfect sense to her. Poor people were housed there because it was polluted, she charges.

The first signs that something was amiss came when one of her sons developed a nasty, persistent rash on the front of his legs and knees where they came into contact with the soil when he knelt to play in the dirt. One day he brought in a two-headed frog and showed it to his mother and kept it in a jar with holes in the lid until it died. "I just thought it was a freak of nature," she recalls.

Bishop, now sixty years old, who lived at the complex for thirty years, also began to notice that her health and that of her children and neighbors went downhill fast after moving to Midway Village. When she arrived at the housing complex, she was holding down two jobs and weighed 170 pounds. Over the following years, she developed a painful scoliosis (curvature) of the spine, ballooned up to 350 pounds, and walked around bent over on crutches. After leaving Midway she had surgery in which a titanium rod was inserted to straighten her spine, and she is now back to her original weight.

Bishop's eldest son was the hardest hit as a consequence of living atop the toxic wastes, she claims. One day he said: "Ma, look at this." He was pointing to his left eye, which had become crossed overnight. From her work as a nursing assistant, Bishop knew that something in his head was pressing on his optic nerve and took him to the hospital. He was operated on immediately, and a small tumor was removed from his brain.

The children in the Midway Housing Complex seemed to suffer the most severe medical problems because they played on contaminated soils, Bishop speculates. Many had eye problems: "They had to get eyeglasses that were as thick as Coke bottles," she says. In addition, many of the children had behavior problems: "A kid would be perfectly normal one minute and then would lose it [become irrational] for no reason." There were also too many children born deformed and with reproductive abnormalities, she continues. Many young female residents were sterile, had fibroids, or suffered from unusual menstrual bleeding that cleared up if they moved away, she adds. "Too many neighbors were sick or dying," she concludes.

For years, Bishop's grandchildren stayed with her at Midway eight to ten hours a day and often overnight while their mother worked. Over time they developed rashes, headaches, sleep apnea, nosebleeds, and a host of other health problems that required a large number of medications. When Bishop and her grandchildren moved out of the complex and into her daughter's house some distance from the site, all of the health problems disappeared, she notes.

## Rude Awakening

The first clear indication that Midway had a contamination problem came in 1990 when Bishop, president of the residents' association, was contacted by the San Mateo Housing Authority and told to convene a meeting of residents. At the meeting they learned that workers would soon arrive to "beautify" Midway Village and install a new drainage system. To Bishop's surprise, when the workers arrived, they were wearing hazmat suits. Asked why they were covered in protective gear, one of the workers replied: "For the same reason that you should go back inside and close the doors and windows."

Bishop was confused and asked: "Well, what about us? Why do you need protective gear and supposedly we don't?" Later she befriended one of the workers, who told her that he had been warned there were hazardous wastes in the ground and that he should keep his suit and mask on at all times. However, when residents asked officials about the contamination, they were told, "It's no more dangerous than it is to eat barbecued chicken where some of the skin has been charred on the grill." Bishop did not buy these assurances. If the housing complex was so safe, she wondered, then why did the cleanup crew have to wear protective gear, why were they pouring concrete patios in front of some of the housing units to cap the toxic soil, and why were residents being told not to plant gardens and instead to use newly installed above-ground planters?

She also felt betrayed. "They suckered us big time. They used me," Bishop says, anger rising in her voice. Prior to the arrival of the workers in what the residents referred to as "spacesuits" Bishop was on the residents' association beautification committee. She was one of a number of residents asked to decide what color to repaint the buildings and where trees and bushes should be planted. "Here I was telling new residents that we were blessed to be in this beautiful place and that we needed to keep up the appearance of the units and make the outside tidy. I was telling them that when all the time the housing was all built on toxic wastes."

Looking at Midway's grounds with new eyes, Bishop noticed areas in the grass smeared with a black, tarry substance. "The kids would get this black stuff on their clothes," she says, and they tracked it indoors. Later, when bulldozers and backhoes started digging up the grass, she could distinguish distinct layers of soil that were heavily contaminated with a black substance. Bishop also claims she saw a barrel containing a black substance that was unearthed during the excavation. "They just cover the stuff up, and then they have to dig it up again seven years later," Bishop says. "I guess you could call it a cover-up."

Asked if drums of toxic wastes had been unearthed on the grounds of the Midway Village Housing Complex, Robert Doss, manager of environmental support services at PG&E, said they had not. "This was a popular urban legend," he observes. What was unearthed was not a drum of hazardous wastes but rather a footing for playground equipment that was approximately the size of a drum. As for a cover-up, Doss is emphatic that there has been none: "That's just wrong. Investigations of the contamination have been carried out in the open. Meetings have been convened in the neighborhood. Documentation has been made available in public places . . . and the regulatory agencies have been very responsive to requests for information from residents."

**Amateur Environmental Investigators**

But this description of how events unfolded does not resonate with Bishop. She and some of the other residents had to educate themselves about the contamination and its health effects and become toxics detectives. They went on a paper chase to find state and federal data about the level of contamination surrounding their homes. Some of these documents were available at the EPA's Region IX headquarters in San Francisco and Berkeley, but as they made repeated requests, their access to documents was restricted, they claim, and copying costs were high. Eventually the truth will come out about the poisons Midway residents are living on, Bishop contends. "You can bury this stuff for years, but eventually it bubbles up. It's like a lie. You can cover it up for awhile, but eventually it will come to the surface. These people lied to us with a straight face. They will boil in hell for it," she prophesizes, concluding, "Someday they will have to realize that people shouldn't live here."

**Mistakes Were Made**

Is Bishop correct in her view of the dangers of contamination at Midway? Mistakes were clearly made by government officials who initially permitted the Midway Village Housing Complex to be constructed on chemically contaminated soils. As early as 1944, federal housing authorities were informed by a building contractor that the land on which they proposed to build Navy housing was contaminated with "much decomposed lamp black and oil refuse mixed with the mud."[1] It is possible that at the time, no one fully appreciated how toxic the substances were or what diseases they might cause. More recently, however, since the contamination of Midway has been quantified using modern monitoring techniques, questions remain about whether residents were informed about the extent of the contamination in a timely manner and about the possible health effects.

Questions also remain as to whether residents should have been relocated rather than leaving them at the housing complex while the polluted earth that surrounded their homes was dug up and hauled away. Time and again, testing at the site revealed hot spots of contamination despite large-scale efforts to remove the problem. Despite these repeated failures to fully clean up the grounds, regulators never moved Midway residents out of harm's way. Officials say relocation was not necessary; residents and activists disagree.

Almost everything about the safety of living at Midway Village Housing Complex is hotly contested. What all parties agree on, however, is that the public housing was built on contaminated fill that came from an old PG&E facility. Also uncontested is the fact that the children of the low-income, largely African American resident population at Midway played in the contaminated soils. It is widely known that many Midway Village residents ate food out of their gardens grown on earth that was later carted away to class I hazardous waste disposal sites where highly toxic wastes are deposited.[2] It is also indisputable that the series of extensive and expensive efforts to clean up the site, which have taken place over the past sixteen years, have made no attempt to analyze or remove contaminated soils from beneath the housing units.

On every other subject, residents and environmental justice activists say one thing—and PG&E officials and state and federal regulators contradict them. For example, residents claim that their health has been adversely affected, while corporate, judicial, and state regulatory officials say that health studies have failed to demonstrate this. Residents claim that a study found high levels of chromosomal abnormalities among residents, while federal officials argue that the study was flawed. Health experts disagree over whether there is an elevated level of disease at Midway. Residents contend that they continue to live on top of Superfund levels of toxic chemicals; regulators say that contaminated soils have been removed and pathways of exposure are now blocked, making the area safe for habitation.

Midway's History

The source of the contamination at Midway Village Housing Complex dates back to the turn of the century when the Manufactured Gas Plant (MGP), a subsidiary of PG&E, operated from 1905 to 1916. The plant left behind a legacy of toxic residues in the soil including tarlike residues and a substance known as lampblack. Lampblack, a finely powdered carbon coal, contains polynuclear aromatic hydrocarbons (PAHs) and volatile organic compounds (VOCs)—both classes of chemicals known to cause a wide variety of illnesses. Cyanide was also found in the soil.[3]

In 1944, in an effort to find a building site for additional Navy housing, the U.S. government invoked its power of eminent domain to appropriate ten acres of PG&E land as well as some adjoining private property. "As part of the construction of housing soil contaminated with PAHs was removed from the former MGP property and used for grading," notes Wilma Subra, who was part of the Midway Village Review Committee tasked with reviewing documents to ascertain whether the cleanup of contaminants at Midway Village was adequate to protect the inhabitants.[4]

This initial mistake, committed by government officials during World War II, which put at risk the health of Navy midshipmen and subsequently residents of Midway Village, has had a far-reaching impact. Jockeying to avoid responsibility for this tragic mistake continues to this day. While the land was under federal ownership PG&E had no control over what was done on it; in addition, Robert Doss, manager of environmental and support services for PG&E, holds that there was no evidence of contamination at the time. This assertion, however, is undercut by a report suggesting that the areas of Midway Village and the small, adjacent playground, Bayshore Park, "that received fill could be visually identified by the characteristically dark color of the Lampblack."[5]

With the wisdom of hindsight, the U.S. government's use of contaminated soil for the grading of a residential housing site appears not just ill advised but also possibly criminally negligent. Lawsuits seeking restitution for affected residents, however, have failed. The suit against the federal government was dismissed because the government does not permit itself to be sued in these kinds of cases. Furthermore, several suits against PG&E, the San Mateo Housing Authority, and regulatory agencies were dismissed with a finding that the litigants had failed to establish that exposure to the contaminated soil had harmed resident health and, some say, because lawyers botched the case.

In 1955 the Navy housing complex constructed on contaminated fill was turned over to San Mateo County for public housing and for schools, and the ten acres that the federal government had confiscated from PG&E were returned to the company. In 1976, the U.S. Department of Housing and Urban Development (HUD) provided San Mateo County with funds to construct Midway Village, a subsidized low-income housing complex with thirty-five multifamily town homes with 150 units on eighteen acres.[6]

Meanwhile, the PG&E site next door to the Midway Housing Complex was transformed into the utility's Martin Service Center, where crews take equipment to be serviced. There was no disposal of hazardous wastes on the site nor were hazardous wastes burned there during the period that the site was used as a service center, Doss claims. This was not a dangerous facility to live next to, he maintains. The facility "has a low hazardous quotient," he states. But residents have a different memory of what was going on across the fenceline from their apartments. Helicopters and trucks would bring large pieces of equipment to the PG&E site, and barrels of materials would burn late into the night emitting acrid odors, residents report. Doss describes this as untrue: no helicopters brought equipment or materials to the site, he states emphatically.

It wasn't until 1980, while regrading their site, that PG&E officials found "significant quantities" of lampblack and MGP residues on their property and reported this fact to regulatory officials. By 1984 the contaminated area was listed by state regulators as a heavily contaminated site that would have been eligible for state Superfund monies to clean it up had PG&E not agreed to pay to haul away the contaminated soils. Much of the hazardous waste on the PG&E site has been trucked to class I hazardous waste sites, and a lengthy berm of other wastes has since been capped with concrete. Unknown at the time was how much of the contamination had already spread across the property line into the residential area.

## Contaminated Soils Discovered

In the process of investigating the extent of pollution at the PG&E property, officials began to test the soil across the fenceline in Midway Village where PAHs were discovered in the soil in 1989. The levels of PAHs on the residential grounds were orders of magnitude below those found on the PG&E grounds (around 10,000 ppm), Doss asserts. And while the levels at Midway Village were above background levels, amounts commonly found in the region, "they were not much more," he adds. PAHs are products of combustion, Dos explains, and are nearly ubiquitous compounds found in car exhaust, soot, and forest fires and on barbecued meat.

While PG&E officials downplay the level of contamination and health risk at Midway Village from exposure to PAHs, contamination levels at the housing project were considerably above the 10 ppm standard established as safe. In fact, some samples taken from the subsoil adjacent to living units were as high as 626 ppm. Other reports suggest that the early sampling of soils at Midway Village found PAH levels that were "150 times the normal level common to urban areas." This is significant, the report continues, because "lab animals exposed to high dosage of the chemicals [PAHs] exhibit a greater incidence of lung cancer, stomach, and skin cancer, birth defects, immune deficiencies, and respiratory ailments."[7]

Having found above-background levels of PAHs at Midway Village, PG&E officials notified the San Mateo Housing Authority, which owned the residential complex; the California Department of Health; and the state's Department of Toxic Substance Control. They did not, however, notify residents at Midway of their findings. "That was the responsibility of the Housing Authority," Doss notes. But a number of residents disagree. PG&E was not forthright with its residential neighbors about a potential health problem, claims LaDonna Williams, a former resident of Midway. In fact, she asserts, PG&E engaged in a deliberate attempt to hide the full extent of the contamination.

Once the toxics at Midway were discovered during the course of efforts to improve the drainage on site, on December 16, 1991, the Department of Toxic Substances Control issued an order requiring PG&E, the San Mateo Housing Authority (SMHA), and HUD to carry out a remedial investigation and cleanup of the contamination. HUD paid 90 percent of the costs of the remediation, PG&E paid 9 percent, and Daly City paid 1 percent.

**Health Impact**

LaDonna Williams grew up a quarter-mile from the Midway Village Housing Complex near the old Navy housing in an area across the freeway from what was then baseball's famed Candlestick Park. As a child, Williams used to play near the abandoned MGP. "There was a little creek down there next to the plant where we used to splash around," she recalls. At the time Williams had no idea that she was mucking about in soils that would later be listed as a Superfund site polluted with arsenic, lead, cyanide, benzene, diesel-range petroleum hydrocarbons, naphthalene, and other PAHs and VOCs. These chemicals of potential concern in Midway's soil and indoor air would subsequently become the subject of multiple investigations by state toxics regulators.

When Williams moved into Midway Housing complex, there was no fence between the residential and park areas and the PG&E facility. There was no warning sign either, she recalls. "My kids would be in the dirt all the time digging tunnels and leaping off the mounds as if they were superman towers," she continues. "And then they would track dirt into the house."

For years Williams had been struck by how many Midway residents were sick. When she took her children to Kaiser Permanente or St. Mary's hospitals with medical problems, she often met other Midway Village residents. Their children were suffering from deformities, seizures, severe neurological conditions, asthma, rashes, hair loss, and other medical problems "you might expect from chemical exposure," Williams claims. Residents of Midway were also experiencing high rates of diabetes and cancer, she adds.

But her first inkling that the housing project was contaminated with toxic chemicals from the PG&E facility came in 1990, a year after Williams moved out of Midway Village, when her mother was diagnosed with cancer. Her suspicions were aroused when a friend told her that "men in spacesuits"—head-to-foot rubberized protective gear with gas masks and goggles—were digging up stuff at Midway Village and people were being told to close their windows and stay in their houses. Could the excavation and removal of soils at Midway Village have something to do with her mother's cancer, she wondered?

Williams now recognizes that there were early signs that the soil in which she grew her vegetables was not healthy but at the time she ignored these omens. "For one thing the worms were weird, and there were two-headed frogs. When the kids would catch the frogs, they'd bring them to us and show us that they had no eyes or were missing limbs," she says. Dogs around the housing complex also looked mangy and scabrous, and the horses in a nearby barn were "skinny, sick looking." Even the trees died, she recalls. Williams now reasons that the invisible toxins that were contaminating the plants and animals were also wreaking havoc with her family's health.

The list of ills her family experienced would discourage even Job. Her mother and stepfather died of cancer when they were fifty-one years old. Both lived nearby and visited Williams's home in Midway Village frequently. They helped with her garden, growing okra and tomatoes, and ate out of her garden. "There is no doubt in my mind what killed them. Because of what's in the ground at that damn place, I had to bury them," she charges.[8]

"This is still a Superfund site," Williams told agency officials during one of the many meetings she attended. "You claim to protect the health of the public. Then do it. You have knowingly placed us on this heavily contaminated land—the dirtiest land in San Mateo County—and you have used it for low-income housing. This is environmental racism and environmental genocide," she charged. In response, she recalls, representatives from both the state and federal EPA told her that she was trying to create a panic in the housing complex. Robert Doss is particularly upset to hear his company accused of environmental racism. "This could not be further from the truth," he says. PG&E has an explicit environmental justice policy in place that is applied throughout its operations. "Statements like this are amazing but not unexpected," he adds.

But is the charge of environmental racism so preposterous? *San Francisco Weekly* reporter Martin Kuz compares the cleanup PAH contamination from a Gas Manufacturing Plant at Midway, where most of the residents are people of color, with the cleanup of similar pollution at Alhambra near Los Angeles, where most of the residents were white. Several differences between the two cleanup initiatives suggest unequal treatment. First, in Alhambra, all residents were relocated for six months while the cleanup took place, whereas at Midway, residents were left in their homes. Second, all the driveways, sidewalks, and patios at Alhambra were removed so that contaminated soil under them could be removed. This did not take place at Midway. Third, an average of four to five feet of soil was removed at Alhambra, while two to five feet were removed at Midway. Fourth, and most telling, soil from the crawl space below the homes in Alhambra was removed, whereas the dirt beneath the living units at Midway remains untouched.[9]

## Getting Organized

But even before these comparisons were made, Midway Village residents found nothing inappropriate about the charge of environmental racism. In 1997, frustrated with the unwillingness of regulatory officials to relocate them, LaDonna Williams, who had moved out after a decade at Midway, and a small group of Midway residents who remained at the complex formed People for Children's Health and Environmental Justice. A core group of this organization still meets periodically and lobbies for relocation and compensation. Among them are Irma Anderson, Mary Tanner, and Maria Downing.

Anderson is a long-term Midway resident activist who has knocked on doors as part of an informal health survey and passed out leaflets warning about the dangers of contamination. Most of the residents who were outspoken have either died or moved away, and those who remain are afraid to speak out because they don't want to lose their housing, she explains. The controversy over the contamination has torn the community apart, and Anderson receives obscene phone calls and has doors slammed in her face by residents worried that activism over contamination will get them all kicked out of public housing.

Despite this occasional harassment, Anderson, who had a nine-pound tumor removed from her stomach nine years after moving to Midway, continues to protest. "I'm just tired of waking up with a headache," says Anderson, who describes living with nosebleeds, rashes, and recurrent cysts during her residence at Midway. When it gets cold, she is afraid to turn on the heat because the furnace will suck more toxins into her home. "I'd love to move, but I can't afford to," she says. Despite many setbacks, Anderson still holds out hope for relocation: "God will touch the hearts of the officials, bring them to the negotiating table, and cause them to treat us as humans regardless of our race." She just hopes she will be moved and compensated before she dies, she adds.

## Soil and Air Sampling

Over a fifteen-year period, soil and air sampling at Midway Village and Bayshore Park has been extensive: nine investigations were launched and some eight hundred samples taken and analyzed. In a paper reviewing the data obtained during this period, Charles Salocks, staff toxicologist at the California Environmental Protection Agency's Office of Health Hazard Assessment, Integrated Risk Assessment Branch, reveals that PAHs were found in many soil samples along with cyanide.

PAHs were found in all 35 soil samples in 1989 and cyanide was found in 5 of the samples. In June, 1990, PAHs were found in 10 of 14 samples taken while digging a drainage trench; PAHs in the soil ranged from near 0 to 72 ppm. (The target cleanup goal for PAHs is 10 ppm.) In 1992, 69 of 70 soil samples contained PAHs in the top two inches of soil; benzene and diesel range petroleum products were also found. In a 2000 test of 426 soil samples, PAHs were found in both surface and underground soils, with the deeper samples showing higher concentrations. Testing in May 1993, during which 169 samples were taken, found 0 to 629 ppm PAHs in subsurface samples, while surface samples ranged from 0.022 ppm to 169 ppm PAHs. In 2001, 9 of 60 samples still exceeded screening levels for PAHs. From 2001 to 2002, high levels of PAHs were found in the floor of excavated areas, and naphthalene, a carcinogen, was found in soils being removed from the site. In 2002 naphthalene was found in the air at the Midway Village Housing Office.[10]

Salocks concedes that there are some PAH concentrations in the soil beneath resident homes and beneath the excavated areas that contain PAH concentrations above 10 ppm target cleanup goals. Opportunities for short-term exposure may still exist, and the depth and breadth of the contamination have not been completely mapped, he adds. Despite these caveats, Salocks concludes that PAH exposure at concentrations above target cleanup goals "has been largely eliminated, and as a consequence any risks to human health have been eliminated as well." A report from the federal Agency for Toxic Substances and Disease Registry comes to a similar conclusion: "[The agency] believes the soil cleanup levels [at Midway Village] to be protective of public health. Therefore, additional sampling, beyond what is being proposed in the near future, does not appear warranted."[11] The bottom line, according to these reports, is that it is safe to live at Midway Village.

## Department of Toxic Substance Control

The state regulatory official most knowledgeable about the contamination at Midway Village is Barbara Cook, branch chief of North Coast

Cleanup at California's Department of Toxic Substance Control. Cook oversees the testing and cleanup of hundreds of sites such as the one at Midway Village where contaminants are discovered close to residential populations.

According to Cook, the pathways of exposure at Midway Village have been blocked. To illustrate her point, she produced a map of every living unit at Midway Village, carefully numbered and labeled. Using the diagram, Cook makes a case that the most intense levels of contamination were in the southern portion of the complex along the fenceline with PG&E. This makes sense, she points out, since the contaminated soil came from the facility. To deal with the higher levels of contamination, several measures were taken. First, residents were offered a chance to move to vacant units elsewhere at the complex while the soil removal was being done. (Only one resident availed herself of this opportunity, she notes.) Plastic coverings were placed over the windows of units near the excavation, and dust suppression measures were used to reduce resident exposure. Then 12,261 cubic yards, or 11,000 tons, of contaminated soil were dug up and removed from around the living units where contamination was discovered. In addition, two feet of topsoil from the 3.8-acre Bayshore Park were removed. To further block possible pathways of exposure, clean fill was brought in to replace the soil that had been removed, and new concrete patios and asphalt walkways and parking areas were laid down. Redwood planter boxes were installed to discourage residents from planting anything in the soil to further reduce the possibility of exposure.

The real problem at Midway, says Cook, is that "there is a huge distrust issue" and some residents just don't believe anything that regulatory officials tell them. "Some of the residents wanted me to provide them with a new house, money for compensation, and payment for medical expenses," she recalls, "and I told them I couldn't give that to them."

## Help from Outside

Unable to convince regulatory officials that they should be relocated and compensated for damage done to their health, a group of residents at Midway Village contacted Wilma Subra, a chemist with extensive experience helping grassroots fenceline groups. Midway residents were fortunate to get help from Subra, who runs a chemistry laboratory and consulting business, the Subra Company, out of a modest office opposite a sugarcane field in New Iberia, Louisiana. With degrees in microbiology and chemistry, Subra was in a position to provide not just technical assistance but also a consultation about how Midway residents could position themselves to win their struggle for compensation and relocation.

After visiting Midway Village and attending meetings with regulatory agencies, Subra delved into the technical data generated by the numerous soil studies, reclamation efforts, and regulatory findings. She then wrote highly technical letters and memos to the various regulatory agencies, and this input made its way into the five-year review deliberations about possible further action at Midway.

"These people [at Midway Village] should be relocated. No one should be made to live on top of contaminated soil," Subra asserts bluntly. To back up this assertion, she notes that the extensive remedial soil removals at Midway did not haul away dirt from beneath the homes of residents and that contaminated soils also remain under sidewalks and streets. This means that contamination may be vaporizing out of the soil beneath the living units and accumulating in the indoor space where residents live. Furthermore, only 10 percent of the "primary contaminated area" at Midway Village was tested, leaving the 90 percent of the area uncharacterized, Subra reports. She is convinced that testing of the surface and subsurface soils on the rest of the Midway Village grounds would find widespread contamination above the 10 ppm cleanup goal for PAHs. "The entire site has PAHs, mostly in excess of Remedial Standards [10 ppm]," she continues. Even areas that have been remediated are problematic because arbitrary limits were placed on the extent of the cleanup, she adds.

For example, in some areas, a decision was made to excavate to a depth of two, three, or five feet. When further sampling was done in the floor of these excavations, PAHs were found in concentrations up to twenty-four times the cleanup standard, she notes. Logic would dictate that further excavation should be done until a depth was reached at which the level of PAHs in the soil met the cleanup goal. Instead, officials decided to stop the excavation and replace the toxic soil with clean fill to block any possible pathways of exposure. Subra finds fault with this decision, which left the possibility that "contaminants could migrate into the clean soils," she points out. A monitoring program should be set up to "track the rate and quantity of movement" of surface and subsurface contaminants, Subra advises.[12] "The health of the community was not protected by such a limited scope [of both testing and cleanup]," she concludes.

Barbara Cook is less worried than Subra about the possibility of PAHs migrating through the clean soils that have been laid down as a cap over the contamination. PAHs do not migrate easily, she points out, and they are not easily soluble in water. As a result, the possibility that PAHs will percolate up from beneath or travel horizontally through the clean fill is remote, she says.

Subra, however, believes there is a danger that PAHs may be vaporizing from the contaminated soils beneath Midway's housing units and outgassing into the homes of residents. Efforts to seal the houses off from the earth beneath them have been less than perfect, notes Subra, who is concerned that during renovations, there was "an apparent lack of a process to seal wires and pipes where they enter homes."[13] This creates a pathway for chemicals in the soil to vaporize and enter resident homes, she says. An ivy vine in the home of Irma Anderson, which has grown up from the soil beneath the house into a crack between the baseboard and her subfloor, suggests that there are holes through which air can leak between the area under the living units and the inhabited space above. "This [the ivy growing through the crack] is a completed pathway of exposure," observes Subra, who calls for more indoor air testing below and in the homes of residents under a variety of weather and seasonal conditions. Ongoing inspections at the Midway Village should also be conducted by county Housing Authority inspectors to ensure that exposure pathways are sealed off, she adds.

## Dearth of Grandparents

As to whether Midway residents have already been harmed by exposure to PAHs, Subra has an informed opinion: "The health of the community is still being impacted by the contaminants on which they live." PAHS "are toxic at very low concentrations." The PAHs and VOCs identified in the air and contaminated soil at Midway are "known neurotoxins and they could be responsible for a wide range of the illnesses that residents are reporting," including problems with children's eyes, light sensitivity, and people going blind; women reporting uterine cancer, infertility, and abnormal menstruation; and chromosomal aberrations and birth defects.

Residents have lived at Midway long enough that the latency periods for cancer and birth defects have already passed, Subra notes. As a result, she anticipates that a comprehensive health survey would likely detect elevated levels of cancer and birth defects. Based on her own observations, Subra thinks the health problems at Midway may be serious. She describes the community "as a neighborhood without grandmothers. . . . The majority of grandparents have died from medical conditions at a fairly young age. The medical problems have been attributed to the toxic chemicals on which they live."

"Who should be held accountable [for health problems resulting from contamination at Midway Village] is a big question," Subra observes. "Should it be the Housing Authority for not warning residents? Should it be PG&E? Should it be the health and environmental agencies for not warning the Housing Authority and the community? There is a long list of parties that could be held responsible." In the end, the smart thing to do may be to move the people out, find them new housing, demolish Midway Village, cap the contamination, and keep people off it, she continues. "But that is a long way off," she concludes.

**Health Survey**

In the absence of state or federally sponsored comprehensive health studies at Midway Village, some residents did their own health survey by going door-to-door in the housing complex. They found what they considered to be high levels of disease and early deaths. Some made their own list of ills they suspect were caused by exposure to toxic chemicals coming out of the PG&E facility and printed it up as a handout sheet: "As a result of this exposure to 300 toxins, many carcinogens, the community suffers numerous illnesses which include cancer, tumors (brain, stomach, breast), respiratory/breathing problems, asthmas, miscarriages, sterility, birth abnormalities/disabilities in children, learning disabilities, skin growths, discolorations, and rashes, chronic bloody noses, neurological disorders, heart abnormalities, digestive disorders, unexplained loss of hair, seizures, and death."[14]

Officials say that these anecdotal reports compiled during a resident-run health survey do not provide objective, scientific data. However, in February 1996, a professional health survey was carried out by Rosemarie Bowler, a professor of psychology at San Francisco State University who also holds a graduate degree in public health. Her study, initiated at the request of the Midway Village Residents Association, was funded by the Boccardo law firm, which represented the residents in a lawsuit. With the help of six graduate students, Bowler's staff contacted 138 of the 153 households at Midway Village. Of the 138 households contacted, 58 completed questionnaires. When compared with a control group of Oakland residents, Midway residents who completed the questionnaire were 6.7 times more likely to have skin rashes and 1.5 times as likely to report acute bronchitis—medical problems that are consistent with exposure to PAHs, Bowler notes.

Midway residents also reported higher levels of anemia, asthma, allergies, and psychiatric disorders than did residents in the control group in Oakland. In a matched-pair analysis, Midway residents reported symptoms in eleven of thirteen categories 1.5 times more often than those in Oakland. Statistically higher levels of symptom reporting were found in seven of thirteen categories: dermatological (6.8 times), respiratory (6.6 times), sensory (6.1 times), gastrointestinal (5.7 times) neurological (4.7 times), headaches/chemical sensitivity (4.5 times), and cardiovascular (2.6 times).[15]

"The results of the study suggest heightened symptoms, illnesses, and medication use for the residents of Midway Village. The findings are consistent with the effects of chemical exposure as shown in previous studies by the principal investigator and in the literature," Bowler writes. There is also a relationship between exposure and symptoms, with those who reported greater exposure generally reporting more symptoms, she continues. "In conclusion, this study suggests an association of PAH exposure and the health status of Midway Village residents and that indeed these residents of Midway Village have been adversely impacted by the presence of toxic chemicals in the soil of Midway Village."[16]

The quality of Bowler's study was called into question by expert testimony provided by Marc Schenker, chair of the University of California at Davis Department of Public Health Sciences, during a subsequent lawsuit. Schenker found many deficiencies in Bowler's study, noting that it was not peer reviewed, had inadequate control groups, failed to examine former residents, and had an inadequate discussion section.

In her defense, Bowler says that Schenker held her health study to an impossibly high standard given that it was carried out on a shoestring budget with a staff of six graduate students. "It was like comparing apples and oranges: this was not a National Institutes of Health epidemiological study with $500,000 in funding," she observes. "It was a small health survey," she says, but it had adequate controls and an acceptable methodology. Bowler says she became discouraged when the judge sided with Schenker in the lawsuit and never ended up publishing the study because she was involved in other work. But it remains a good study and at the very least provides enough evidence of adverse health impacts to suggest that a more comprehensive health survey is warranted, she adds.

Asked if it was possible that Midway residents were experiencing disproportionately high levels of disease, PG&E spokesperson Doss says that while he is no medical expert, every medical expert who studied the residents say that "this is not the case. Everything I know leads me to believe that the area is safe and no additional efforts are needed."

## Lawsuits

Two lawsuits were brought by residents who wanted to be compensated for harms allegedly done to them by exposure to toxic chemicals at the Midway Housing complex. Some 250 residents of Midway joined a class action lawsuit against the federal government for endangering their health and asked for $1.25 million in compensation. It was dismissed on the grounds of federal immunity.

The second lawsuit began in July 1991 when fifty-five residents of Midway sued the Housing Authority, San Mateo Housing Authority, and PG&E for failing to protect their health. According to Doss, who was deposed during the course of the five-and-a-half year lawsuit, some twenty-five residents were examined at San Francisco Hospital, the University of California Medical Center, and the University of California at San Francisco. At these facilities, residents were given physicals by medical examiners, and the results of these examinations were reviewed by occupational health experts, epidemiologists, and learning specialists. These experts found "that no one had been injured by exposure to the relatively low levels" of PAHs, Doss claims.

In 1994, the judge dismissed the lawsuit for lack of what he saw as credible evidence proving that residents had been harmed by contamination at Midway. The case never went to trial. The judge ruled that residents had failed to provide concrete evidence that their illnesses were caused by pollution from the PG&E facility. Another lawsuit was dismissed in August 1997 by a judge in San Mateo County Superior Court.[17] The case was appealed and the ruling upheld.

As far as Doss is concerned, that ended the legal dispute. However, despite having won a dismissal of the suit, PG&E offered cash settlements to a number of residents. Some took them, and others didn't. Residents report the settlements were between a few hundred and a few thousand dollars. "We were not paying for injuries because it was not established that there were any," Doss explains. But the company did pay undisclosed amounts for what Doss terms "psychological stress." The settlements were offered because "people had been subjected to anxiety. This is a hot button issue" where people felt threatened in their own homes, he adds.

Having lost the lawsuit, a group of Midway residents turned to state and federal regulators for relief. "They gave us the runaround," former resident Lula Bishop reports. The Housing Authority would point to PG&E, and PG&E would point to the Navy and the Department of Toxic Substance Control. No one wanted to take responsibility, she continues. "So many people became sick and disabled that we wanted not just to be moved to other public housing, we wanted to be compensated for the damage done to us," she continues. "We wanted lifetime medical and financial aid and housing. Those were our demands . . . but it never happened." Stymied once again, LaDonna Williams wrote a letter to President Bill Clinton, who ordered that Midway Village be reinvestigated by federal regulators. A new study was done of the soil, and a cleanup was done of the child care center and park but not under the housing units, she notes.

### Greenaction

In some ways Midway is worse than other fenceline communities located adjacent to obvious chemical hazards, observes Bradley Angel, executive director of Greenaction, a Bay Area environmental justice group. "It is rare that you come into a community where people come up to you—a complete stranger—and lift up their shirt or roll up their pant legs to show you their rashes," he says. "The situation in Midway was absolutely outrageous in terms of corporate environmental racism and widespread disease among the residents," he continues. "These people were living on top of a toxic time bomb. There were Superfund-level, cancer-causing chemicals here, and no testing was done of the soils under these people's homes or under the day care center."

Angel has worked with grassroots fenceline groups exposed to chemicals from nearby plants for over twenty years. For eleven years he was Southwest toxics coordinator for GreenPeace USA. When GreenPeace directors decided to get out of grassroots environmental justice organizing, Angel and a number of his colleagues founded Greenaction to carry on the work, and Midway was one of their first major campaigns. To Angel, who had worked in a series of low-income and heavily minority communities located next to hazardous waste sites and heavily polluting industries, the situation at Midway was familiar.

State regulatory officials maintain that they have removed the contamination through excavation and remedial work, he observes, "but we say they have removed only some of the problem and that people are still getting sick. This is a state Superfund site, and every time they do testing of the soil, they find more hot spots. That should tell them that there is a serious problem still remaining. We argue that residents should be moved out and compensated," he asserts.

To make this point, Angel and others at Greenaction worked with Midway residents to organize at least ten civil disobedience actions where they blocked PG&E trucks that were hauling contaminated soil away from the grounds of the Midway Housing complex. They argued that it was unsafe to excavate toxic wastes while residents were on site and therefore exposed to dust from the removal operations. And they contended that no amount of excavation would make the site safe for residents.

At one point, state and federal regulatory officials promised residents and Greenaction activists that they would conduct testing underneath the subsidized housing units and the day care center, but they never made good on their promise, Angel claims. As a result of this and other lies, there is no trust among residents about official efforts to clean up the site. "It is both racist and reckless endangerment to place families over soils contaminated with Superfund levels of chemicals. I think it is also criminal," Angel continues. "This was a toxic cover-up," he adds. "Midway Housing complex has to be closed. No one should live there. Government officials should not be gambling with these people's lives," he concludes. Although much of the protest has died down at Midway, Angel thinks that eventually residents will win their struggle. "The final chapter on Midway has not yet been written," he predicts.

## Toxic Trap: Public Housing near Heavy Industry

It is not unusual to find public housing built immediately adjacent to refineries, chemical plants, and public utilities that handle large volumes of toxic chemicals. No one should have to live immediately adjacent to heavily polluting plants, but it is particularly odious when building permits are issued that permit the poorest and most vulnerable members of society to be housed in public complexes where the air, water, or soil is contaminated by a nearby industrial plant. Yet in Daly City, California; Port Arthur, Texas; Pensacola, Florida; Norco, Louisiana; and a host of other fenceline communities, this is exactly what occurs.

Another now-famous case in which public housing was dangerously located near heavy sources of pollution can be found on Chicago's South Side, where the Altgeld Garden public housing complex, designed to house ten thousand residents, was built in 1945 to house veterans returning from the battlefields of World War II. The complex was built on an industrial waste dump and surrounded by fifty-three highly toxic industries, including oil refineries, cement plants, steel mills, coke ovens, and incinerators. The location of the complex was so close to so many industries that it came to be known among residents as "the toxic donut." Some 97 percent of the residents in Altgeld Gardens are African American, and 62 percent of them are below the poverty line.[18]

The reason that public housing complexes such as these are built immediately across the street from refineries, chemical plants, and military bases is no mystery: land is cheap near high-emission facilities, and there is little organized resistance to building low-income facilities in these locations. The result of this kind of callous calculation is predictable:

low-income people who find shelter in public housing often have a host of ailments (or susceptibility to illness) that is exacerbated when they are trapped in highly polluted housing complexes. Headaches, nosebleeds, skin rashes, neurological problems, retardation, asthma, respiratory disease, reproductive disorders, and cancer are often found at elevated levels in these dangerously located, government-subsidized properties.

Why do residents not just leave if conditions are so bad? Most stay in these toxic traps because they fear losing their berth in public housing and ending up living in shelters or on the street while waiting their turn on long lists for the next available unit. Placing public housing residents on the fenceline with toxic industries sends a powerful message that society does not value these people and is willing to sacrifice their health in order to reduce public spending. Not only is this type of decision morally repugnant, it is also shortsighted in that taxpayers ultimately pay for emergency health care services for this population once they fall ill from foreseeable chemical exposures.

President Clinton got it right when he said that no child should have to grow up next to a hazardous waste site. He issued an executive order to all federal government agencies to review their procedures from an environmental justice perspective. Following the logic of this executive order, residents in public housing projects built on the fenceline with heavy industry, such as those at Midway, should be relocated to safer neighborhoods.

### Eskimos Exposed to Toxic Military Wastes

The following chapter turns to another group of Americans who have been disproportionately exposed to toxic chemicals after hosting a U.S. military base. The problem first came to light when Annie Alowa, a Yupik Eskimo midwife living in a community on Saint Lawrence Island in the Bering Sea, noticed a precipitous increase in cancer among her people. The U.S. military had set up bases on the island and then dumped toxic chemicals in the ground before withdrawing. Alowa was suspicious that the contaminants had entered the food chain on which her people depended for their sustenance.

Subsequent environmental investigations revealed substantial contamination of the soil and water beneath two of the largest villages on the island and a favored hunting, fishing, and gathering camp. As residents became aware of the scope of the contamination, they organized protests and collaborated with an environmental group from the mainland. As a result of these protests and the media coverage of the problem, the government was pressured into conducting a massive, costly, and, ultimately, incomplete cleanup of the affected areas. No apologies or compensation have been extended by the government to any of the victims of this military contamination.

# 10

## St. Lawrence Island, Alaska: Yupik Eskimos Face Contaminated Water and Traditional Food Supplies near Former U.S. Military Bases

The lesson to be drawn from the experience of the Alaskan Yupik Eskimos who live on St. Lawrence Island in the remote and ice-laden Bering Sea is simple: no one is beyond the reach of pollution's long arm. This is particularly true in communities that played host to U.S. military bases.

Were she alive today, Annie Alowa could testify to this sad fact. Before her death from liver cancer, Alowa lived in one of the most remote and once-pristine areas in the United States. A Yupik Eskimo and resident of Savoonga and Northeast Cape, Alowa's home was on a biologically rich promontory of St. Lawrence Island, Alaska, one of the western-most outposts of the United States. Isolated in the Bering Sea near the Arctic Circle, St. Lawrence Island is 135 miles southwest of Nome, the dog-sled capital of Alaska, and 36 miles across the water from the Chukchi Peninsula in Russia.

Alowa's extended family live in both Northeast Cape and Savoonga, a village founded by reindeer herders, comprising rough-hewn wooden homes (elevated above ground level to accommodate abundant snows) along with some newer modular houses. Her people practice subsistence hunting of walrus, seal, whale, and reindeer; fishing of salmon, halibut, and trout; trapping of fox; shooting of birds; and the gathering of wild greens and berries. In this fashion the Yupik people have inhabited St. Lawrence Island for over twenty-five hundred years.

From her front door, Alowa could see no giant refineries spewing smoke or chemical plants leaking an alphabet soup of esoteric and hazardous chemicals. In short, her home was positioned to be one of the least polluted sites in the nation. It remained so until the U.S. military set up radar bases on the island to give early warning of Japanese forces during World War II and, subsequently, Russian aircraft and missiles during the Cold War.

## Military Outpost

St. Lawrence Island became strategically important during World War II when the Japanese attacked the nearby Aleutian Islands, capturing Attu and Kiska. During the war, many of the men of St. Lawrence Island enlisted in the Alaska Territorial Guard—"The Eyes and Ears of the North"—and guarded the land against a possible Japanese invasion. With the advent of the Cold War, residents of St. Lawrence Island were recruited to form patrols guarding against infiltrating Russian frogmen. Cold War tensions ran high: the island was buzzed by Russian MiG jet fighters, a blackout was enforced, and dog sled teams were staked out around the villages to warn of nighttime intruders. For its part, the U.S. military erected and manned an aircraft control and warning system radar station on the island to detect Russian warplanes. Subsequently when a Russian missile attack became a serious concern, the island played host to sixty-foot-tall parabolic radar antennas that were a component in the Alaska Integrated Communications Electronics system ("White Alice").[1] The U.S. Air Force operated one of these radar bases at Northeast Cape from 1958 to 1972 (it began construction in 1954) as part of the North American Air Defense Command.

Erecting these radar bases on an Alaskan island ringed by ice most of the year was a monumental logistical challenge. At the village of Gambell and later at the encampment at Northeast Cape, barges landed during the summer months, and bulldozers hauled ashore the large diesel generators and transformers that powered the array of military antennas. The Northeast Cape Air Force Communications base, which residents helped build, was large in size compared with native encampments, encompassing forty-eight hundred acres. In addition to the huge radar arrays, there were twenty-five large industrial buildings and associated support facilities.[2]

The arrival of the U.S. Army at Gambell in 1951 and subsequently the U.S. Air Force at Northeast Cape in 1954 was initially greeted by local residents as an opportunity to contribute to the national defense effort.

Staunchly patriotic, the Yupik people have enlisted in the U.S. armed services, the National Guard, and the Alaska Territorial Guard in large numbers. Hosting a pair of military bases on the island also provided a number of cash-paying jobs (rare in their subsistence economy) building and subsequently servicing the bases.

Seen from the perspective of the thousands of years that St. Lawrence Island has been inhabited by the Yupik people, the period during which U.S. military personnel were stationed there—lasting only from 1951 to 1957 in Gambell and from 1954 to 1972 at Northeast Cape—was a mere blink of the eye. However, the toxic legacy left behind by these military operations proved lasting. In the thirty-five years since U.S. soldiers and airmen left the island, the land has been found to be contaminated with over 220,000 gallons of diesel fuel, as well as solvents, polychlorinated biphenyls (PCBs), dioxins and furans, lead-based paints, asbestos, heavy metals, and unexploded ordnance including 30- and 50-caliber ammunition and grenades. Deliberately or accidentally spilled onto the ground or dumped into unlined pits, some of these toxic chemicals made their way into the food chain that local people depend on and into the waters from which they drink and fish.

## Health Impact

It was Annie Alowa, a village health aide, who first noticed that serious illnesses were increasing among Savoonga residents who frequently visited Northeast Cape. The increase in disease coincided with the closing of the military base sixty-five miles southeast of Savoonga, she observed. In 1998 Alowa calculated that out of the one hundred residents who were regular visitors to Northeast Cape, fourteen neighbors (including herself) developed cancer. This explosion in cancer rates was new, charged Alowa, who had overseen the health care of her people for twenty-five years. The only logical explanation was that they had been affected by the military contamination, she said. Alowa took her concerns to the U.S. Army Corps of Engineers, the Alaska Department of Public Health, the Alaska Department of Environmental Conservation, and the U.S. Environmental Protection Agency. For her trouble, she was sent on an endless merry-go-round of regulatory visits that yielded no concrete plans to clean up the mess at Northeast Cape.

Frustrated by this bureaucratic runaround, Alowa contacted Pam Miller, now at Alaska Community Action on Toxics (ACAT), who was investigating the health impacts of several of the more than six hundred formerly used defense sites (FUDS) in Alaska, the majority located in areas inhabited by Native people. By December 1998, Alowa was in the terminal stage of liver cancer when Miller interviewed her on camera and produced a documentary of her struggle, *I Will Fight Until I Melt*. Since then, ACAT has been collecting blood and soil samples, which document that residents of Northeast Cape have elevated levels of PCBs in their bodies.

Since Alowa's death, a growing number of St. Lawrence Island residents have been speaking out about the pollution on their island and the health consequences they think occurred as a result of the contamination. One of them is her daughter, Chris Alowa, now fifty-nine years old, a retired elementary school teacher who lived with her mother at Northeast Cape and worked briefly at the base. "I'm mad at the military for not telling us about the danger from the contamination. I swam in the Suqi [Suqitughneq] River, our family drank from those waters, and we ate fish caught in them," continues Alowa, who was raised at Northeast Cape and whose mother died of liver cancer, father died of intestinal cancer, and has had her own bout with breast cancer. "A lot of the people who lived at Northeast Cape with us have since died of cancer," she observes. "The government owes us a lot. They think we are just like guinea pigs for an experiment. They polluted our best subsistence hunting and fishing place—a place I keep in my heart—and it has left a big hole," she adds.

Since Annie Alowa's early protests, the perils of the contamination at Northeast Cape have become better known, and many islanders have been frightened away. As a result, a community of about a hundred has dwindled to about ten summer residents. "As more individuals become sick with cancer, others are increasingly afraid to engage in subsistence food gathering activities in areas [such as Northeast Cape] they have utilized for many generations," notes a report from Coming Clean, a nonprofit that focuses on health impacts of chemicals released from industry and the military.[3]

## Large Volume of Toxic Chemical Releases

One of the most outspoken critics of what he describes as an "inadequate and superficial" cleanup effort at Northeast Cape mounted by the U.S. Army Corps of Engineers is forty-eight-year-old Delbert Pungowiyi, president of the Native village of Savoonga, the tribal government. Pungowiyi and his colleagues successfully led a grassroots campaign to thwart military plans to create a monofill—a large pit in which military debris would be buried at Northeast Cape. The Department of Defense (DOD) attempted to make the monofill attractive to islanders by offering them a financial incentive in the form of a "tipping fee" for every load of debris buried at the dump. A local coalition opposed to the monofill overcame this powerful cash inducement and unified opposition to the plan, forcing the DOD contractor to abandon this strategy. Instead, at considerably greater expense, the department agreed to ship large quantities of debris and contaminated soil from the demolished base to licensed landfill and hazardous waste facilities in Washington and Oregon.

Despite a cleanup effort at Northeast Cape, which has cost the government $48 million, Pungowiyi remains unconvinced that the source of pollution has been rooted out. Large areas of contamination that remain below the surface are gradually leaking more poisons into the environment, he contends. The largest spill occurred in 1969, when a bulldozer blade punctured a fuel tank, releasing an estimated 160,000 gallons of diesel fuel at Northeast Cape. This was not the only spill. A pipeline break at Cargo Beach caused by a bad weld released 500 gallons of diesel fuel.[4] There were also releases at the main pumping station. causing some observers to estimate that that 220,000 gallons of diesel spilled into the ground and the Suqi River that drains into an estuary.[5]

The diesel spill and PCB contamination created a localized ecological disaster in one of the richest hunting, fishing, and gathering grounds for the Yupik. Northeast Cape was a prized area because so many important food sources were concentrated in a relatively compact area. The Suqi River ran so thick with fish that it was child's play to catch them, a seal rookery and haul-out point made them easy game, and abundant edible greens and berries along with a source of freshwater made the area an attractive place to live year-round. The area was also an important resupply point for residents of Savoonga who were headed out to other camps as well as a place to weather out bad storms.

## Yellow Seals

In this hot spot of biological diversity, the massive diesel spill burned a hole in the ecological web like a cigarette through silk. The spill caused some seals to turn yellow—the color of diesel fuel. "You could smell the diesel miles before you arrived at the village," Pungowiyi recalls. "The seals and shore ice and tundra were soaked in diesel. The spill wiped out the seals and fish [although some have since returned], and a number of birds including cormorants died," he continues. The pollution forced natives to travel to other locations to find game and fresh drinking water. Even today, a multicolored petroleum sheen appears in the river water, he adds. PCB-laden dust also drifted onto the greens and berries that residents gathered for food.

As a result of exposure to this contamination, the level of PCBs in the blood of Savoonga residents who frequent Northeast Cape is higher than other Savoonga and Gambell residents and ten times higher than average levels among U.S. residents, Pungowiyi notes. Many of the people who lived year round or transited through the area occupied by the former base suffered health problems as a result of exposure to toxic chemicals left behind by the military, he continues. As an example, Pungowiyi points to the fact that his adopted father, Clarence Pungowiyi, who worked at the base and never smoked, died of lung and liver cancer. "Before the military came, there were fewer deaths that baffled us," he observes. "Fewer people died of cancer and many residents lived to advanced ages. Now there are multiple cases of cancer diagnosed every year that were unheard of in the past."

Pungowiyi wants the U.S. government to provide restitution to Native residents for the value of the lands lost due to the military contamination. He also wants the government to recognize the significant contribution residents made to the Cold War effort. "Something is terribly wrong here, and we are having a hard time getting anyone to pay attention to the problem," says Pungowiyi. To draw attention to their cause, tribal leaders are considering mounting "a journey of grievance" trip to Washington, D.C., to put their case before Congress.

For confirmation that a large quantity of below-ground contamination remains to be cleaned up at Northeast Cape, Delbert Pungowiyi need only turn to his cousin, Perry Pungowiyi, a whaling boat captain whose record take of three whales in a single season in 2007 fed much of the village. From 2000 to 2004, Perry Pungowiyi was hired to run heavy equipment by Nugget Construction and Bristol Environmental, which were under contract with the Army Corps of Engineers to clean up Northeast Cape.

At first Perry Pungowiyi felt good about working on the cleanup because he was part of a team removing large amounts of hazardous materials before shipping them off the island. But as the work progressed, he began to get concerned. When he extracted partially buried barrels out of the ground, he could see numerous others beneath them, but he was forbidden to dig up the buried barrels. "That's not in our contract," he was told. The job was only to clean up the debris on the surface, not anything that was buried. "We would bulldoze soil over the barrels and flatten the area out, and they [his supervisors] would call it good," he adds. "It hurt my feelings. I felt helpless. I was not allowed to really clean up the contamination." Everybody who worked there knew there was considerable subsurface contamination, he continues. "We weren't really there to clean it up. It was just to give the place a facelift and make it look good."

Another practice that he felt caused further contamination of the soil was the digging of a pit surrounded by an earthen berm that was then lined with plastic. Old fifty-five-gallon drums, some of them half-filled with liquids, were rolled into the ditch and then squashed with an excavator. (Army Corps of Engineers officials deny any knowledge of this practice.) In the process, tears in the lining were created that allowed contaminants to leak into the soil. Perry Pungowiyi was appalled: "We are the ones who live off the land. If the food chain is contaminated, it hurts us because we are part of that chain."

### "The Ocean Is My Garden"

Perry Pungowiyi is by no means the only resident who has direct personal knowledge of the contamination at Northeast Cape. One of the most respected elders in Savoonga, seventy-eight-year-old Alexander Akeya and Annie Alowa's brother, who worked briefly at the base as a cook, was witness to one of the early oil spills there. "I saw it with my own eyes," he says, and claims he watched army personnel empty the fuel in a lengthy pipeline into the estuary. After the oil spill, the birds and fish had a bad smell, but residents kept hunting and fishing. Akeya was deeply troubled by the deliberate emptying of fuel into the estuary: "The ocean is my garden . . . and they spoiled my garden. I eat from the sea," explains Akeya, who lost both his brother and sister to cancer. "There was nowhere to holler it up." At the time there was a deep reluctance to complain about the behavior of military personnel who caused contamination and nowhere close by to report the problem.

Many residents were surprised at both the extent of the military contamination and the fact that the military would abandon its bases on the island without cleaning them up. "The Army left while we were sleeping," observes Fritz Waghiyi, a fifty-five-year-old tribal council member from Savoonga. As a boy, Waghiyi remembers fishing salmon and gathering berries at Northeast Cape where his father did seasonal work at the base. "We didn't know until they left that there was any danger to us from the contamination," he says. "I think my family's health was affected by the contamination. I lost quite a few relatives from cancer and think there is a connection. I had uncles and aunts who lived there who died in their middle to older years of cancer." He believes the military caused environmental damage and should restore the natural resources. A hunter, fisher, whaler, and electrician by trade, Waghiyi is frustrated by the slow pace of the cleanup and the fact that many contaminants are being left underground. He also feels that there has been a lack of adequate consultation between military authorities involved in the cleanup and the tribes.

Not only did the military pollute Northeast Cape, the Department of Defense also failed to warn local people about the dangers of contamination there, notes fifty-seven-year-old Merrill Annogiyuk, who visited the Native settlement for three years while his father worked at the base. "The government could have at least informed people that it [Northeast Cape] was off limits until it was cleaned up and people would have stayed away," he observes. Annogiyuk remembers seeing transformers and barrels broken open, with their contents spilled into the soil. "I was playing around in the lagoons where the diesel leaked. I guess I've got some chemicals in my body now. They took some samples of our blood and sent us information about our levels," adds Annogiyuk, who died in January 2008, six months after he spoke these words.

Annogiyuk also recalled his father salvaging lumber from the base to enlarge their home. "There was black paint on the lumber that we later learned had lead in it and the lumber smelled when it got hot," says Annogiyuk, whose brother died of cancer. "We used to check the dump for things we could use. There were fifty-five-gallon barrels and two-by-four lumber as well as spoons, forks, knives, dishes, and other things that were hard to come by. . . . Things were hard at that time, and people were poor here. Some people found food at the dump that wasn't spoiled, and they took it because it was tasty." Later testing would find that some of the areas where Natives salvaged construction materials were contaminated with PCBs, petroleum residues, asbestos, and heavy metals.

## Keeping the Anger Inside

Sylvia Toolie, the forty-nine-year-old tribal coordinator at the Village Council in Savoonga, also worked at Northeast Cape, and a number of members of her family still live there. "We are a pretty naive people, and at first we didn't understand the problems posed by contamination," she observes. There was one large spill of diesel fuel into the Suqi River that caused the fish to die, she recalls. Then the military pulled out of the base, but her family and a number of other residents stayed on unaware of the danger

Looking back, Toolie thinks that the contamination may have affected her family. Her father died of lung and liver cancer at age fifty-one, an uncle died of a neck tumor at age forty-nine, and she contracted breast cancer, which is in remission following treatment. The men who salvaged materials and worked on the base seemed to be hit the hardest, Toolie observes. Some of those who worked there year round developed premature bald spots on their heads, she recalls. "It was unusual: we never used to have so many deaths from cancer at these young ages," she adds. Toolie worries about the health of her aunt and uncle who continue to live seasonally at Northeast Cape and still drink the water and eat fish caught there. Some of the game that local residents depend on for their subsistence has been contaminated, she continues: "Sometimes the insides of the seals and walrus are kind of green, spoiled, smell bad, and cannot be eaten."

While much of the surface debris and grossly contaminated soils have been removed, Toolie says a lot more needs to be done. "I'm kind of angry about what happened," she says. "In fact, I'm very angry [about the contamination], but it is my nature to keep this anger inside. We are taught to respect our elders and show respect to anyone who comes from outside." The tribal leadership has told residents to wait until the Army Corps of Engineers completes the cleanup before debating whether to demand reparations for the damage done to their hunting grounds. But Toolie discerns a change in the view of the elders: "They are starting to realize how bad the contamination is at Northeast Cape and, hopefully, they will ask for reparations."

When the military came to the island in the 1950s, they signed an agreement with the governing tribal council committing the Department of Defense to clean up the area when they left, says Vi Waghiyi, who was born and raised in Savoonga and is now an ACAT organizer based in Anchorage. The agreement is quite specific, she continues: "Any refuse or garbage will not be dumped in streams or near the beach within the proposed [base] area, as this will prove detrimental to the seal breeding grounds."[6] Clearly the military failed to abide by this agreement, she notes. "The military is not above the law and needs to be held accountable."

### Extent of Contamination

A large, complex, protracted and expensive cleanup effort, directed by the U.S. Army Corps of Engineers, was undertaken in 1984 under the Defense Environmental Restoration Program. The man in charge of the cleanup at Northeast Cape and Gambell, Carey Cossaboom, is a geologist hired by the Army Corps of Engineers to work on the cleanup of eight FUDS in Alaska. These include sites in both Gambell and Northeast Cape, the latter of which has become one of the program's most expensive cleanup effort to date.

The cleanup job Cossaboom faces is substantial, technically difficult, costly, and controversial. At the height of the military occupation at Northeast Cape, there were 212 soldiers and airmen erecting and operating a radar base in a remote island where the weather conditions are severe. The military facilities included barracks, a mess hall, dayrooms, a motor pool, a cinema, and a bowling alley, among other amenities. A landing strip was constructed for resupply of the base, and a tramway was built to haul supplies up Kangukhsam Mountain to the radar dome site. Multiple transformers, containing highly toxic PCBs, were needed to operate the radar system, which required enough power daily to supply the equivalent of twenty-five thousand homes.[7]

When the bases on St. Lawrence Island were closed, almost everything was buried in Gambell and abandoned at Northeast Cape. The military materials left behind included "containers of hazardous substances such as brake fluid; fuel drums containing petroleum products; anti-freeze; above and underground fuel tanks; and containers and transformers containing PCBs."[8] According to one account, "one barrel dump contained more than twenty-nine thousand drums some leaking unknown fluids."[9]

Pollution at Northeast Cape became the focus of an environmental investigation in 1985 and passed through four investigative and remedial phases. Much of the base's physical facilities, surface debris, and a considerable amount of contaminated soils were containerized and hauled off the island by contractors (Bristol Environmental and Nugget Construction) working for the Army Corps of Engineers from 1999 to 2005.[10] To date 17,000 tons of waste and debris and 315 tons of PCB-contaminated soils have been removed, Cossaboom calculates.

But more work remains to be done: half of the PCB-contaminated soils have yet to be removed, he continues. Another request for $14 million has been proposed to dig up primarily diesel fuel–contaminated soils and remove the rest of the PCBs, but the money has yet to be appropriated, he adds. There is also some contamination in the sediment of the Suqi River and in wetland areas that will not be removed. Instead these areas will be monitored. Residents make clear that they will continue to push for the removal of contaminants in both the Suqi River and the estuary at Northeast Cape.

Elevated Exposure

A review of Army Corps of Engineer documents reveals the extent of the mess left behind by the Air Force. For example, the area where the transformers were housed in the heat and electrical power building area in the main operations complex was found to be heavily contaminated: 141 tons of PCB-containing soils had to be removed. Benzene and naphthalene were detected in the soils. The puncture of a 400,000-gallon diesel fuel tank during routine snow removal is estimated to have released 160,000 gallons of diesel fuel, contaminating the grounds up to a depth of sixteen feet. Sampling found 17,000 and 26,500 ppm of diesel range organics (DRO, while the cleanup standard is 9,200 ppm).[11] Ground-water under and near the complex was found to be contaminated with DRO, benzene, and lead to a depth of ten to twenty-five feet.

Sampling at the closed military base during 1994, 1998, 2002, and 2004 showed the concentrations of DRO, benzene, and lead in the groundwater had gradually diminished due to natural attenuation, but contamination continues to exceed cleanup-level standards. One of the lessons here is that if the military studies a contamination problem over a ten-year period in an area where deep snow falls and then melts off in the spring, it will face a less extensive cleanup job at the end of the study period because the runoff will have flushed much of the contamination into the sea. Although this reduces the tax dollars spent on the cleanup, this approach is not a bargain for local residents who drink the water and eat seals, whales, walrus, and fish that ingest pollution from the runoff.

In July 2007, the Army Corps issued an end-phase cleanup strategy entitled "Proposed Plan for Northeast Cape Air Force Station."[12] The plan calls for two more years of work during which "limited excavation" and removal of contaminated soils would take place in some areas. For example, at the main operations complex, the Corps plans to remove some six thousand cubic yards of contaminated soils down to a depth of five feet. After that the site will be monitored and five-year regulatory reviews performed. In most of the thirty-four contaminated sites, the Corps recommends "no further action," arguing that these areas have been adequately cleaned up. Land use controls will be issued in some areas, such as a temporary restriction on the installation of a drinking water well.

## PCBs in Blood of Residents

A controversy over the danger that PCB contamination poses to St Lawrence residents broke out between officials at the State Division of Epidemiology and two experts from the State University of New York (SUNY), Albany and Oswego campuses, who were hired as research team members by ACAT.

Recognizing that more facts were required in order to determine if contaminants left behind by the military at Northeast Cape posed a continuing health threat to island residents, Pamela Miller director of ACAT, applied for and was awarded a grant from the National Institute for

Environmental Health Sciences to assess the body burden of chemicals in the blood of local residents of St. Lawrence Island. ACAT research team members mounted a collaborative effort that included residents of Gambell and Savoonga, the State University of New York School of Public Health and the Environmental Research Center at Oswego, and Norton Sound Health Corporation, the health provider for St. Lawrence Island residents. David Carpenter at SUNY's School of Public Health and Ron Scrudato, director of the Environmental Research Center and Superfund Basic Research Program also at SUNY, were hired by ACAT as research team members. Both are experts on environmental contamination and have worked with Native American populations, including the Mohawk communities in New York State.

As part of the initiative, four village health aides who were island residents were trained to take blood and environmental samples and conduct health interviews. They took sixty blood samples: twenty from Gambell and forty from Savoonga. Twenty of the forty residents chosen for testing in Savoonga "used Northeast Cape regularly." An analysis of the results revealed that the sixty islanders tested had an average of 7.7 ppb of PCBs in their system, or what Carpenter described as five to ten times the PCB body burden of an average person in the lower forty-eight states, which is 0.9 to 1.5 ppb. The highest levels were found among residents who frequented Northeast Cape.[13]

"These results show significant PCB and persistent pesticide contamination of the Saint Lawrence Island Yupik people," Carpenter observes. "While some portion of these contaminants derives from atmospheric transport of contamination, our results show a greater elevation of PCBs in the blood of those individuals who used Northeast Cape for traditional or occupational purposes. We conclude that the PCB contamination from the formerly used military site at Northeast Cape resulted in increased human exposure." He found further evidence of "ongoing exposure" to PCBs in several people with camps at Northeast Cape.

While it is true that the Arctic is a sink for PCBs and other persistent organic pollutants that accumulate in cold environments and concentrate in the fat and tissue of animals and people, there is also evidence of a more localized source of PCB contamination at Northeast Cape, the ACAT report continues: "Improper disposal of PCB-contaminated transformers and other materials at the military site at Northeast Cape on St. Lawrence Island has resulted in elevated levels of PCBs found in soils, water, plants, and fish. Exposure to people may occur through ingestion of contaminated foods and water, through the skin, or inhalation. PCBs cause adverse health effects at exceedingly low levels."

PCBs are more dangerous to human health than previously recognized, Carpenter continues, and can cause a variety of cancers, immune system suppression, irreversible effects on brain development and IQ in infants, and disruption of the endocrine system. "Studies have documented health effects in people with comparable levels of PCBs as those found in the Saint Lawrence Island Yupik people," he adds.[14] Carpenter's colleague, Ron Scrudato, is concerned that the lack of comprehensive characterization of the pollution at Northeast Cape and the designation of many contaminated areas there as requiring "no further action" means that St. Lawrence Island "will continue to serve as long term source of contaminants to the arctic region."[15]

Localized exposure to PCBs of Eskimo populations that depend on subsistence lifestyles is not unique to St. Lawrence Island residents. The Arctic Monitoring and Assessment Report estimates that thirty tons of PCBs were used in the sixty-three military radar stations that were set up along the 66th parallel in Alaska, Canada, and Greenland. Unknown amounts of PCBs were also disposed of in landfills at these sites.[16]

### Benefits of Traditional Diet Outweigh Risks

State officials were quick to disagree with ACAT's conclusions that the military bases on St. Lawrence Island had resulted in dangerously high levels of PCB levels in the blood of some residents. Officials at the state Division of Public Health, Section of Epidemiology argued that the PCB levels found in St. Lawrence islanders "are similar to other Alaska Native populations." The report went on to fault the SUNY study for failing to factor in age differences, which are important because older people tend to have higher concentrations of PCBs in their blood. They also noted a failure to take into account the difference in PCB levels between populations that depend on traditional food hunting and gathering as contrasted with those who buy food at a store.

Authors of the state report concluded that the levels of PCBs in the blood of residents from St. Lawrence were similar to those of other Alaska Native populations exposed to PCBs through global transport of the pollutants through the food chain—and not from a localized source such as the military bases on the island. Although the authors made clear that they supported the cleanup of toxics on the island, they nevertheless felt that the PCB levels detected in the blood of residents "are unlikely to cause adverse health effects" and concluded that "no adverse health effects would be expected from PCB concentrations measured in this study." They recommended that residents continue their consumption of subsistence foods, which, they argued, provide benefits that "far outweigh the controversial potential adverse health effects from contaminants at the concentrations found in these foods."[17]

Miller remains skeptical that the problem is not serious. Both military and heavy-industry officials have "perceived Alaska as a remote, sparsely populated place with a weak regulatory structure. Neither has felt much obligation to clean up after itself or take responsibility for the damage that is being done."[18] State and federal officials frequently dismiss reports of contamination problems as "anecdotal" and without an adequate scientific basis, she observes. Often the U.S. Department of Defense and the Department of Energy downplay the stories of people being made ill by military contamination by arguing that in fact the problem is with the lifestyles of the people involved, who may be falling ill due to excessive smoking or drinking.

Another controversy concerns whether the waters of the Suqi River are safe to drink. State and federal regulatory officials insist that the water is safe to drink, while the ACAT research team contends it is not. The lower Suqi River contains PCBs, PAHs, and mirex, a highly toxic chemical banned in the United States and used as a flame retardant and pesticide, notes Vi Waghiyi, coordinator of ACAT's St. Lawrence Island Environmental Health and Justice Project.[19] State officials claim the river water is safe to drink. "But our data contest that finding," insists Pam Miller. "Saying that the river water is safe to drink is criminal and shows a sheer disregard for local residents." Cossaboom at the Army Corps of Engineers disagrees: "Contamination of the Suqi River and estuary sediments and surface water does not exceed cleanup levels."

## Gambell

Northeast Cape is not the only area on St. Lawrence Island polluted by the U.S. Department of Defense. St. Lawrence Island, known by its natives as Sivuqaq, is ninety miles long and twenty-two miles broad at its widest point. At the northwestern end of the island lies the ancient settlement of Gambell, a community of some seven hundred Yupik residents. Gambell is a great place to live for those who make their living fishing, collecting bird eggs, and hunting whales, walrus, and seals. The village is located on a spit of gravel between a mountain visited by large numbers of birds, a lake rich in fish, and a stretch of ocean where seal, whales, and walrus appear seasonally. Abundant marine mammals and seabirds are drawn to this deep-water part of the Bering Sea by the nutrient-rich Anadyr current.

Many of Gambell's residents are the descendants of survivors who made it through the famines of 1878 and 1880 that decimated the island's previous population of some four thousand. The famines occurred after islanders encountered outsiders arriving from the lower forty-eight states on whaling ships. This early cultural contact proved disastrous for the Yupik, who were infected first with smallpox and later with tuberculosis. A further plague came in the form of alcohol brought in by archeologists who pillaged old sites for artifacts and paid the Yupik men with alcohol, residents report. The alcohol so debilitated village hunters that, in 1878, they failed to take advantage of the summer months when much of the subsistence hunting, fishing, and gathering normally takes place. As a result, many villagers perished from starvation; only 264 were said to have survived. Since then the total island population has grown to over fifteen hundred (down from a high of about four thousand) as a result of improved nutrition and health care. In addition, orphan children from Nome have been adopted by families in Savoonga and Gambell to widen the gene pool.

During the Cold War, three branches of the U.S. armed forces—the Army, Air Force, and Navy—all established a presence at Gambell. The first Army troops came ashore in 1948 to establish a base camp, the Air Force built a radar surveillance and intelligence-gathering base in 1951, and the Navy later laid underwater communication cables. All of these bases were abandoned by 1956. But in its day, Gambell was on one of the front lines of the Cold War. Russian MiG jet fighters buzzed the island, and Gambell residents were asked to join a ground observer corps. Some eighty of the male residents enlisted in the Alaska Territorial Guard and pulled three-hour shifts over a twenty-four-hour period to watch the skies and beaches for Russian infiltrators. Volunteers were issued a Springfield bolt-action rifle, ammunition, and a uniform. They kept up this around-the-clock guard duty until 1957. The Cold War turned briefly hot on June 22, 1955, when two Russian MiGs shot down a Navy P2V-5 Neptune surveillance airplane that crash-landed eight miles down the coast from Gambell. At the time, Bruce Boolonwon, a twenty-year-old corporal in the Alaska Territorial Guard, was one of a number of recruits who set off in walrus skin boats to rescue the eight-man crew, one of whom was badly burned.

## All-Terrain Vehicles and Snowmobiles

Today Gambell is abuzz with all-terrain vehicles (ATVs) and, in the winter, snowmobiles that are ridden by men, women, and children—often standing stiff-legged and leaning over the handlebars. In earlier days, however, before the advent of the air strip, food stamps, social security, and fuel supplies, the people of Gambell lived during the cold months in underground sod homes heated with seal oil lamps.

Gambell is now a small coastal village of modest wooden homes, a small cluster of recently built subsidized housing units, and a single grocery store. The school is the most modern building in town. The village boasts a lodge that rents out rooms to travelers, regulatory officials, construction workers, and the bird enthusiasts who flock to the island to view the 5 million birds that nest in the rugged cliffs that loom over town. Most birders make the trip to Gambell to identify and add to their life list some of the exotic Asian species that make the short flight across from Russia.

It was in the Sivuqaq Lodge that I met Winnie James Sr., who was a young man when the Air Force came ashore in 1951 to set up a radar station and is now eighty-four years old. James remembers the U.S. Army camp well because it was constructed close to town. The first soldiers came ashore in 1948 and built an encampment that included twelve large huts for the men, kitchens, dining halls, dayrooms, and motor pools. There was also a tramway used for hauling supplies up to the top of the Sevuokuk Mountain that rises precipitously from sea level a half-mile from the village center. It was atop this mountain that the U.S. Air Force manned an aircraft control and warning station radar dome. Then all of a sudden in 1956, the soldiers abandoned the base, taking with them only their rucksacks and their rifles and leaving everything else buried beneath the ground.

### "People Can Do a Lot of Oddball Things"

"People can do a lot of oddball things," says James, who was thirty-four when he was hired by U.S. Army officers to run a bulldozer. He was instructed to dig pits so that equipment left behind after the base closed could be buried. Fifty-pound sacks of flour were slit open and dumped into the pits: "There was a storm of flour. It was a real shame because that could have fed a lot of hungry kids," he recalls. There is some speculation that American forces destroyed their mechanized equipment, broke up the generators and transformers, deliberately spilled diesel fuel onto the gravel of the village, and deep-sixed their ammunition in the lake to ensure that if the Russians ever invaded, they would find nothing useful left on the island. In effect, it was the equivalent of a scorched-earth strategy.

Whatever the rationale, James was directed to bulldoze pits into which generators and transformer were buried along with large caches of ammunition. At one point, soldiers blasted a hole through the ice covering Troutman Lake, pulled two sled loads of ammunition and grenades out on the ice, and dumped it in the lake, he remembers. His final assignment was to dig a pit large enough to bury the bulldozer he had been operating. "Then the staff sergeant burned all the paperwork," he observes.

After the military pulled out, local residents helped themselves to the building materials that had been left behind. "There were no signs up saying that there was any danger," says James who says he witnessed oil and PCBs from the transformers spilled into the ground.[20] "Some think that we don't understand what was going on, but people here want to see the contamination cleaned up. . . . I'd like to see them dig it out before I die," he adds. As to whether he thinks that the contamination has caused health problems, James speculated that it "probably caused some cancer." People are dying at a younger age from cancers because the chemicals are getting into the water and the game, he opines. "The government should compensate us and dig it out," he adds.

James has little esteem for the off-island contractors flown in to do the cleanup work. "I go out with them and put a stick in the ground to show them where to dig, but they just dig somewhere else," he claims.

However, Carey Cossaboom says that the areas James indicated have been thoroughly searched without result by metal detection equipment. They have also used a backhoe to dig in some of the areas James indicated but have found nothing, he notes. Pam Miller at ACAT takes another view: "Denying people's memories about where the wastes are buried shows a level of disrespect that they [the Army Corps] could not get away with in a white community or one with more political clout."

Clearly James is right in saying that military equipment was buried at a number of sites in and around the village. Evidence of this can be found throughout the town. Piled between the village and the beach are caches of barrels and military supplies that have been dug out of the cobble. (Cossaboom says that this debris has been containerized and shipped off island to a landfill, but at the time I was in Gambell, the debris was still visible.) At one location a rusted bulldozer and fifty-five-gallon drums stand as evidence that James's recollections are accurate. Large amounts of debris have been removed from the area, but clearly more remains to be done. Some of the buried debris appears to be beneath or adjacent to recently erected affordable housing units, and more debris was found during excavation for expanding the school.

The fact that residents came into personal contact with some of the contaminated residue from the Army base comes from the daughter of eighty-one-year-old Conrad Oozeva, who as a child played with a chalky white substance she later identified as asbestos. "We used it for the white 'frosting' on our mud cakes," she recalls. Bathing in the lake was also perilous: "When I was in the water I could feel the bullets under my feet," she says. Army Corps of Engineers investigators failed to find much in the way of buried arms caches, but they did provide an educational program for residents "in case a resident does find ordnance or explosive wastes."

## Pack It In, Pack It Out

"We were taught that whatever you packed into a camp, you had to pack out," says Oozeva, a veteran of the Alaska Territorial Guard who started hunting at the age of ten and was a whaling captain of a boat manned by his sons and relatives. At Gambell, the U.S. military failed to follow that motto, he observes: "They left a real mess." Oozeva claims he witnessed Army personnel using rock hammers to pierce drums of oil and let the contents spew out into the gravel on which the village of Gambell is built. Later, water from the village's well began to taste of oil, and a new one had to be dug in a more distant location near the foot of the mountain, he notes. A number of Oozeva's relatives have had cancer, including his sister and a brother. "The military tried to clean up, but a lot is still in the ground. It will never be our land the way it was before the Army came," he concludes.

"We *are* the living evidence of the contamination," says Jerome Apatiki, sixty-three years old, referring to the PCBs and other contaminants in the blood of residents. A veteran of the National Guard who describes himself as a traditional hunter, Apatiki recalls when the soldiers arrived and started handing out candy and soft drinks once a month. "We welcomed them at first. We didn't know that in the future they would do something bad," he continues. Apatiki is concerned that there is contamination in the gravel under the homes of many residents in Gambell. "At the motor pool, the soldiers used to pour the oil and everything into the gravel, and we didn't know at the time that it would cause a problem," he recalls. Money wasn't a problem when the military decided to build a base in Gambell, "but now that it comes to cleaning up the mess they left behind, they say they lack the money to do it," he observes.

Money does play a role in how many former military sites can be cleaned up and the extent of the reclamation efforts. The U.S. Army Corps of Engineers estimates that it will cost $1 billion to clean up (or "close out") the approximately 160 contamination projects at eighty FUDS in Alaska. Nowhere near that sum has been appropriated. In 2007 Alaska received $20.8 million (down from $50 million in 1998) for its FUDS cleanup program, which was 8 percent of the total appropriation of $262.8 million that Congress made available for the cleanup of the nine thousand FUDS nationwide.[21] To date, $7 million has been spent on the cleanup at Gambell compared with $48 million at Northeast Cape, a much larger site, notes Cossaboom.

## More Than They Wanted to Know

Another resident who is unhappy with the extent of the military cleanup effort is Morgan Apatiki Sr., fifty-six years old and the former liaison between the local residents and the Army Corps of Engineers. Apatiki claims that his job with the Corps was terminated "after I gave them more information [about where contamination was buried] than they wanted." Cossaboom denies that this is accurate.

Military officials are trying to downplay the levels of contaminants that still exist in the soil in order to check Gambell off their list of sites that need further cleanup, Apatiki charges. Too many signs shows that the cleanup at Gambell is far from complete, he continues. Divers looking for highly valued fossilized ivory in Lake Troutman found cases of live ammunition; bore samples in the soil found areas saturated with oil just above the permafrost; the area where the motor pool used to be located is heavily contaminated with oil; a number of transformers that were buried have yet to be found and removed; and a series of 75 foot by 25 foot landfills that stretch from the school to the North Beach have yet to be mapped and inventoried. Given an inadequate cleanup and toxic chemicals remaining in the ground, human health is likely to be affected, Apatiki asserts: "We are drinking contaminated water and inhaling volatile gases that are causing all kinds of sickness including cancer, leukemia, strokes, and all kinds of problems that we never had before."

Aaron Iworrigan, who works as the village innkeeper and previously served two tours of duty with the Navy in Vietnam, sees the military contamination problem as a simple matter of bad manners. If Yupik Eskimos stay overnight in a shelter when they are passing through on their way to their camp, they sweep the cabin and chop wood before they leave the next day so that the next occupant will find it clean and well supplied. "That is the principle that the military should have followed here, but they didn't," says Iworrigan. "They didn't show respect to the land or to the people." Iworrigan is amazed by the wastefulness of the soldiers who pushed useful equipment off a cliff to deliberately destroy it or buried it instead of leaving it for villagers to use. His main concern, however, is a suspicion that the town water supply may be polluted. "The government should compensate us for their use of the land and the damage they did to it," he adds.

## Cleanup Efforts at Gambell

Contamination investigations at Gambell carried out by both federal and state officials have continued for over twenty years. A major contamination study was conducted in 1994, sampling followed in 1996, and in 1997 the Army Corps of Engineers hired Montgomery Watson to clean up visible surface debris. Subsequently, in 1999, Oil Spill Consultants removed 26.8 tons of hazardous wastes and nonhazardous wastes "such as asphalt drums, generators, batteries, empty drums, and transformer carcasses," as well as 71 tons of exposed metal debris and 72 tons of contaminated soils.

Today cleanup in Gambell continues under the direct supervision of the local workforce, Cossaboom explains. Additional funds have been provided for this purpose by the Native American Lands Environmental Mitigation Program (NALEMP), a Department of Defense project that provides funds for the cleanup of debris that is not categorized as hazardous by the larger FUDS program or where cleanup is postponed more than five years due to funding limitations. The Army Corps encouraged the Native village of Gambell to take control of the project because it will allow more local people to be hired and will give them ownership and control of the process. "The Corps believes this program should quell some criticism of the cleanup efforts because the NALEMP crew is tasked with identifying old [military waste] burial sites," he adds.

Following the remedial action that lasted from 1994 to 2003, U.S. Army Corps of Engineers project managers decided that their work at Gambell was done, save in two areas that required further limited excavation. The two remaining problematic sites contained soils with arsenic levels in excess of 11 ppm, lead in excess of 400 ppm, and chromium greater than 26 ppm. This pollution was detected in the soils at the former Military Power Facility and Nayvaghaq Lakes disposal sites. The Corps contractors removed fourteen tons of contaminated soils at these sites, as well as sixty tons of exposed metal debris.[22]

In a number of other areas where debris was detected underground, the Corps decided not to dig it out. For example, unfiltered groundwater at a military landfill near the village high school and subsidized housing was found to be contaminated with trace amounts of beryllium, cadmium, chromium, and lead that exceeded screening levels. The site was considered an unlikely source of drinking water, and so the Army Corps did not excavate.

### "An Unspeakable Injustice"

Annie Alowa was stonewalled in 1997 when she first alerted Army Corps of Engineers officials to the serious pollution at Northeast Cape and the high rates among residents of cancer, miscarriage, low birth weight, and endocrine disorders, charges Pam Miller, director of ACAT, who since then has assiduously kept track of the military cleanup on St. Lawrence Island.

Miller calls the manner in which the Army Corps dealt with Alowa's concerns "disgraceful. Colonel Sheldon Jahn, who was, at the time, the Army Corps of Engineers official charged with investigating the contamination, was very disrespectful to Alowa and told her that the cancers were the result of lifestyle choices such as smoking and drinking and not because of contamination." Subsequently Alowa and Miller teamed up to try to bring attention to the problem posed by military contamination. The next year, in 1998, Alowa was back with news that Army Corps of Engineer contractors had done some sampling of the soil but had refused to reveal to residents the results. "They were keeping it all a secret," Miller recalls.

Miller judges the cleanup on St. Lawrence Island as less than comprehensive. Only surface debris was removed because anything below ground was "not in their scope of work," according to contractors. The fact that the Corps has recently recommended no further action in some of the most contaminated areas is "absolutely unacceptable to many residents. They want it properly cleaned up," she says. The Corps, believes Miller, just wants to check Northeast Cape site off their to-do list.

Backing up her skepticism about the completeness of the cleanup is a U.S. Government Accountability Office (GAO) report critical of the U.S. Army Corps of Engineers for failing to adequately investigate chemical and munitions hazards at 1,468 formerly used defense sites nationwide.[23] "Clearly, as this report shows, the Corps' slipshod investigative work cannot be trusted to protect the health and well-being of our environment or of our citizens," observes Congressman John Dingell. The GAO report found that the Corps lacked sufficient information to support "no further action" determinations in 33 percent of the sites investigated in Alaska.[24]

While important sources of contamination remain to be cleaned up on the island, Miller notes limited but important victories won by local residents and their allies. First, pressure from residents forced the Army Corps of Engineers to create a Restoration Advisory Board to give residents a voice in the cleanup. Second, the monofill option, which would have dumped large quantities of debris (some of it contaminated with hazardous substances) into a single, unlined pit, was rejected despite efforts to make it financially attractive.[25] Third, remedial efforts at Northeast Cape have gone from one of the Army Corps' lowest priorities to one of their highest.

The contamination at Northeast Cape, one of the worst of the 603 FUDS sites in Alaska, should have been declare a Superfund site and "remains a direct threat to the health of residents," Miller observes. Since the source of the contamination remains buried, wastes continue to leak into the river and into the web of life on which Eskimos depend with each spring thaw.

It is inexcusable not only that the pollution occurred but also that the military has never issued an apology to the people of St. Lawrence Island for the massive oil spills and improperly buried toxic wastes, Miller opines, adding that a natural resources damages case is still possible. "I think this is coming. The government behaved in a criminal way," she continues. The despoiling of the rich hunting and fishing grounds of Northeast Cape was "an unspeakable injustice," she adds. Local people want the area properly cleaned up, but it may be impossible to completely repair it, she observes. "This is a human rights issue now and we [at ACAT] are committed to seeing that justice is done."

**Corps Perspective**

Carey Cossaboom sees it differently. He concedes that the military's abandonment procedures when they left the island back in the 1950s were "pathetic," but he argues that the military's carelessness was ignorant, not malicious. Furthermore, he notes, some of the claims about the health impacts of this contamination "seem to be inflated." Cossaboom is skeptical of ACAT's motives: "I don't believe their interests are entirely altruistic. They depend upon grants for much of their funding. And how do you get grants? Sensationalize the issues," he says. The real story is more complicated, he continues. "So many civilization changes were introduced to Native communities around World War II that it seems impossible to isolate a single societal culprit for perceived injustices. I'd bet that life expectancies have not declined with the advent of the military presence on St. Lawrence Island." [26]

The existence of PCBs in the blood of Northeast Cape residents "is not anomalous" when compared with other Alaskans and Canadian Arctic Natives who depend on traditional subsistence hunting and fishing, Cossaboom argues. "I fear that the relentless blaming of the military for all these ills may be diverting the focus from the real culprit, whatever that may be. I would like to see the island's health woes studied scientifically, but that is not the Corps' mission. Intuitively it seems unrealistic that the communications facilities at Northeast Cape and Gambell could cause such havoc. They are not the industrial complexes that you have in the lower forty-eight [states]. These were short-lived, smallish outposts that suffered an ugly spill or two and conducted some poor environmental procedures out of ignorance and bizarre policy. I doubt there was any maliciousness."

Cossaboom appears dismayed at the level of controversy surrounding how comprehensive and effective the cleanup he oversees has been. "I clean up spills. It feels good to clean a place up. . . . I thought I was riding in on a white horse," he says. He contends that the Army Corps has done a good job of cleaning up military contamination on a remote island under difficult climactic conditions.

The Army Corps' work in Gambell is essentially done, Cossaboom continues: "I don't think there are any remaining health threats from the military in Gambell. I believe that the Northeast Cape will have minimal health threats. Our intention is to remove the remaining PCBs that constitute an environmental risk; if PCBs remain in a landfill, they will be buried beneath an impervious cap." There will be some land use controls left in place that prohibit drilling for drinking water on the former base. Summing up the impact of the Army Corps' cleanup at Northeast Cape, Cossaboom says that once it is done, "it will yield an environment that is significantly mended but not fully healed."

Some of the Corps' work on St. Lawrence Island has been innovative, Cossaboom continues, noting an effort to make the cleanup acceptable to the local population. "We initiated a NALEMP project in Gambell and persuaded the IRA [the tribal government] to manage it themselves, offering much-needed assistance along the way," he notes. The Army Corps also offered a monofill option for local burial of inert debris instead of shipping the waste to a landfill in the lower forty-eight states. The money saved by not shipping out these wastes could have provided funds to build proper health clinics and better water systems for islanders, argues Cossaboom. "But," he adds, "outsiders convinced them otherwise."

Some $48 million has already been spent on the cleanup at Northeast Cape, and at least $23 million more is slated to complete the job. Landfill removal could cost another $80 million, Cossaboom estimates. Should more money be spent here, given that larger populations in other areas have been exposed to similar or more toxic contamination? This is the kind of question that involves the harsh calculus that regulatory officials regularly face: Given that there is only so much money to abate pollution problems, where is that money best spent? Understandably Yupik residents of St. Lawrence Island maintain that the government created the problem and therefore should clean it up whatever the cost. Alternatively, the U.S. government could admit that it was too costly to clean the area up completely, post the grounds as a contaminated area, and compensate the native population for the damage done.

### "We Are Not Your Foes"

Over the years, Cossaboom has gotten an earful of criticism about the cleanup efforts from St. Lawrence Island residents. One resident who took a diplomatic approach with him is Harriet Penayah, a seventy-five-year-old midwife who has a number of relatives who worked at Northeast Cape. At a recent public meeting in Savoonga, Penayah told a Corps official: "We are not your foes; we are your friends. We just want to clean up the PCBs that are killing people. . . . We just want you to help clean up the bad things and clean up Northeast Cape so that we can live a healthy life there and eat the fish and drink the water." Unfortunately, the decision makers are far away, and, she observes, no one seems to be listening. Nevertheless, Penayah held out hope that the need for a thorough cleanup would be heeded "by the big people in Washington" and that Yupik residents would be able to live without fear for their children and children's children.

## "Not a Bunch of Dumb Eskimos"

Among those who are least impressed by the progress of the cleanup at Gambell and Northeast Cape is Vi Waghiyi, ACAT coordinator of the St. Lawrence Island Environmental Health and Justice Project. A former resident of Savoonga, who lived there until the age of eight and whose father was the postmaster, Waghiyi now lives in Anchorage, where she is an organizer and spokesperson for ACAT. She retains close ties to the island where three of her brothers live, including Fritz Waghiyi, a tribal council member in Savoonga. Her father died of colon cancer, and her mother had surgery to remove a colon cancer, a disease that was unheard of in her family prior to the U.S. military coming to the island, she says.

"This [military contamination] is a clear case of trespass. . . . We know what is going on here. We are not a bunch of dumb Eskimos," Waghiyi continues, quoting the words of Annie Alowa. "The military has caused impacts that are devastating to our land and environment, that affect our traditional subsistence lifestyle and culture." Island residents are now concerned that their families are suffering from a number of illnesses "that were not apparent until after the military occupation," she reports. Among these are cancers, diabetes, reproductive problems, thyroid disease, nervous and immune system disorders, learning disabilities, and other health problems. It is Waghiyi's sad task to meet many of her fellow islanders when they come to Anchorage for cancer and other serious disease treatments. She visits them in the hospital and tries to see that they have everything that they need.

"We believe the science of the Army Corps of Engineers is biased [and], in some cases, inconclusive by design," she told the National Environmental Justice Advisory Council. The EPA's decision to rank Northeast Cape as a highly toxic site but not place it on the National Priority List of "Superfund" sites "was a political decision," she adds.[27] Alaskan state officials are "not doing a good enough job to ensure a responsible cleanup," she continues.[28] "With this in mind, the U.S. Department of Defense needs to be held liable for their legacy and conduct an adequate and responsible complete cleanup of the two FUDS on St. Lawrence Island," she adds. There should also be restitution for the damage done to the environment and health of residents, she concludes.

## Widespread Military Contamination on Native Lands

The experience of the Yupik on St. Lawrence Island is not unique. Many Native Americans live in remote areas that have been used as sites for military bases and for the testing of weapons systems. For example, the experience of the Metlakatla Indian Community on the Annette Island Indian Reserve in southeastern Alaska has many challenges in common with those that the residents of St. Lawrence Island face. Both communities have contamination problems stemming from the construction and operation of U.S. military bases and from extensive soil contamination, oil spills, PCBs, and lead. Federal agencies have made a commitment to work with this community to address the contamination problem. The area has also been designated a national Brownfields Showcase Community.[29] Elsewhere, extensive uranium mining and processing has also taken place on or near Native lands.

For decades this poisoning of Native lands went largely unnoticed by the rest of the world. Only recently has the extent of lands contaminated by the military begun to be charted. The toll of illness and death among Native people caused by this military contamination is still largely unknown. While considerable efforts and funds have since been used to clean up the mess left behind, no one from the government has ever apologized for the damage done to Yupik lands or people, and no residents have been compensated for their losses. This can best be described as yet another case of environmental injustice.

At the other end of the continent, in Greenpoint, a Polish American community in Brooklyn, New York, local residents are also living atop a large volume of toxic petroleum wastes that seeped into the ground many years ago. Residents learned about the extent of the oil spill lurking beneath their feet only in 2002 when an environmental group spotted a giant oil slick in Newtown Creek, which forms part of the border between the boroughs of Brooklyn and Queens. Residents have long complained of a strong petrochemical odor and unexplained illness, but their complaints had remained unaddressed until they organized, sued Exxon-Mobil, and demanded compensation and an accelerated cleanup program.

# 11

## Greenpoint, New York: Giant Oil Spill Spreads beneath Brooklyn Neighborhood

The ground in Greenpoint, a Polish American neighborhood on the waterfront in north Brooklyn, feels solid enough. But lurking beneath the surface—like an alien substance in a bad science-fiction movie—is a giant viscous blob of oil that soaked into the ground over the course of a century. The oil has been described by one observer as having the consistency of a giant tub of black mayonnaise.[1]

The size of the oil spill, estimated at 17 to 30 million gallons, is thought to be the greatest environmental disaster ever to strike New York City. It measures some twenty-five feet thick in some areas, floats atop the Brooklyn aquifer, and has spread beneath fifty-five acres of industrial, commercial, and residential properties. "The blob" is one of the largest oil spills in American history and is one-and-a-half times larger than the amount of oil spilled by the *Exxon Valdez* in Prince William Sound, Alaska, in 1989. In that incident 11 million gallons of oil gushed out of the supertanker onto Bligh Reef, killing untold numbers of fish, birds, and sea mammals. In Brooklyn, though, more than the fish and blue crabs in the creek are in peril.

Some Greenpoint residents are concerned that flammable methane vapors and carcinogenic benzene fumes from this underground petrochemical lake are leaching up through the soils into their backyards and basements and causing cancers and other health problems. The U.S. Environmental Protection Agency plans to study the site; the New York attorney general's office and Riverkeeper, an environmental group, are suing the responsible oil companies; the state Department of Environmental Conservation is conducting indoor air vapor studies; private tort lawsuits have been filed; and community organizations are turning up the heat on officials to carry out a comprehensive cleanup and health survey.

Spilling a supertanker-and-a-half of oil is not the work of a day: it took over a hundred years for the oil to leak out of giant storage tanks that lined Newtown Creek in Greenpoint. One large release of oil is thought to have occurred in 1919 when a fire burned down a Standard Oil refinery holding 110 million gallons of oil. Some of the oil went up in smoke, while the rest seeped into the ground. Other less spectacular oil spills occurred during the routine transport and processing of oil products over many years.[2]

**Laboratory of Urban Chemical Archeology**

Greenpoint has not always been heavily polluted, and the waters of Newtown Creek have not always been what the Metropolitan Waterfront Alliance describes as "a laboratory of urban chemical archeology."[3] When European colonization of the area began in the seventeenth century, Dutch residents found the marshy area to be rich in oysters, bluefish, and striped bass. Located on a creek that connects to the East River and Long Island Sound, Greenpoint soon became a trading post for fur pelts and other goods. Subsequently the area turned to agriculture as apple orchards were planted and tidal mills were built to grind grain.[4]

This quiet agrarian scene was transformed in 1861 when the Long Island Rail Road established its Queens Terminal at Greenpoint, making it an ideal location for heavy industries such as fertilizer, dye, and chemical plants, as well as refineries and copper smelters. Starting in the 1890s, some fifty small refineries were built along the banks of Newtown Creek in Greenpoint. The bulk of these were purchased and assembled into the mammoth Brooklyn Refinery by John D. Rockefeller's Standard Oil, the predecessor of Mobil, which was subsequently purchased by Exxon in 1999. Other refineries were owned by Paragon Oil, a subsidiary of Texaco, which is now Chevron. In those days, there were few laws requiring oil companies to build containment systems to prevent spills from soaking into the ground and creek, and common practice was to dump oil wastes that could not be sold onto the ground or into the creek.[5]

All of these industries contributed fuel oil, degraded gasoline, naphtha, benzene, perchloroethylene, trichloroethylene, and a host of other chemicals to the great pool of toxic substances collecting beneath Greenpoint.

For generations, the oil quietly seeped into the ground with little official notice. Residents, however, were aware of the problem. As early as the 1890s a "smelling committee" was organized and began patrolling the neighborhood sniffing out hot spots of pollution. Occasional regulatory citations were issued against the offending plants, but no big investigation was ever launched.

Most of the leaking oil had found a place to hide: as water was pumped out of the ground for industrial, commercial, and residential use, the oil filled in the vacuum created by the extracted groundwater. This continued until 1949, when the groundwater supplies were so depleted and contaminated that fresh water had to be piped in from the Catskills. Over the next twenty years, the water table gradually rose, pushing toward the surface the huge reservoir of oil that floated on top of it. Pushed from beneath by hydraulic pressure (and with nowhere else to go), the oil began to seep out of the banks along Newtown Creek.[6]

The first modern alarm over the oil spill was raised in 1978 when a Coast Guard helicopter operator spotted and reported a giant oil plume in the water off Greenpoint. A follow-up Coast Guard investigation in 1979 revealed an immense subterranean pool of oil. This report effectively blew the lid off industry's dirty little secret in Greenpoint. The Coast Guard subsequently installed recovery sump pumps that collected 100,000 gallons of oil.[7]

When the state took over supervision of the cleanup operations, it worked with Amoco Oil (BP) and Paragon Oil (Chevron) to try to contain oil seeping into the river. Then in 1989, Mobil Oil (soon to merge with Exxon) took responsibility for the spill and began recovery operations of its own in 1990.[8] A quiet deal was worked out between state regulatory officials and executives of oil companies that were responsible for the pollution: the oil companies would gradually pump and treat groundwater to remove oil, and the government in return would neither fine them nor require a timetable for the cleanup. By keeping the deal quiet, the state avoided panicking residents about possible damage to their health and property values. In the process, however, state officials failed to live up to their responsibility to provide residents with crucial information that would have permitted them to make up their own minds about how best to protect themselves.

## Rude Awakening

The unpublicized deal between the oil industry and state regulators began to unravel in 2002 when Basil Seggos, chief investigator for Riverkeeper, an environmental watchdog nonprofit, motored into Newtown Creek off Greenpoint. Seggos was on patrol in Riverkeeper's converted lobster boat, used to monitor water quality in the Hudson River and take officials and reporters on tours of toxic hot spots. On October 24, 2002, Seggos, the boat's captain, and a staffer from the offices of Brooklyn city councilman David Yassky, chair of the Brooklyn Waterfront Committee, were poking around Brooklyn's waterways when they struck oil.

"We wanted to mark the spots where the waters were polluted and where residents were engaged in subsistence fishing," recalls Seggos, a thirty-three-year-old attorney who did his apprenticeship at the Pace University environmental litigation clinic. "A lot of low-income people fish in very contaminated areas. They are more concerned with putting food on the table that evening than they are with potential health problems associated with eating contaminated fish," he explains. In an effort to reduce this kind of toxic exposure, Seggos was surveying the area to see where signs should be posted warning of contamination problems.

But what Seggos found that day was more serious than an area that required some warning signs. Off the prow of Riverkeeper's pollution patrol boat he spotted a section of Newtown Creek two hundred yards long and a hundred yards wide that was coated in oil. This was not just a thin petrochemical sheen that creates rainbow hues in the water; it was a quarter- to a half-inch-thick coat of oil floating off Greenpoint. Seggos took photographs of the oil and contacted the state Department of Environmental Conservation (DEC), where officials told him they were aware of the problem. Seggos was unsatisfied with this response. If officials were aware of the problem, why wasn't the oil being vacuumed up? Where were the booms to contain the slick and the rest of the cleanup crew?

Unable to find out why such a big oil spill was not being addressed, Riverkeeper asked to see all the regulatory files concerning the spill under a Freedom of Information Act request. This tactic bore fruit: two months later Seggos had access to twenty boxes of regulatory agency files on the spill that went back over thirty years and detailed the 17 million gallon spill. During the year he spent researching the problem, Seggos found that in some areas, the oil bulged out beneath residential neighborhoods. The depth of the contamination varied by location: in some places, it was close to the surface, and in other areas. it had seeped at least forty feet deep into the ground.

## Getting Organized

Seggos was not alone in his concern about the oil under Greenpoint. For decades a number of community residents had noted the petrochemical sheen on the water in the creek and smelled a strong petrochemical odor. They had urged regulatory officials to clean up the oil mess. When their entreaties went ignored, they organized themselves into the Concerned Citizens of Greenpoint. One of those who took a leadership role in this grassroots group was Irene Klementowicz who, with her late husband, ran a funeral home on Manhattan Avenue in Greenpoint.

Since the 1980s, Klementowicz and five women from the predominantly Polish neighborhood grew concerned that Greenpoint was becoming a dumping ground for highly polluting industries and was being ignored by regulators. "First they spilled all the oil in the ground, and then they put up an incinerator and shipped in wastes from Manhattan, Queens, and Brooklyn as well as medical wastes from Staten Island. But we got the incinerator shut down, so we have had our successes," she notes with some pride. Nevertheless, the oil problem persisted. "The oil is still seeping into the creek. You can see it and smell it in the air. One time when they did some digging for a foundation, I looked at the soil and it was black and oily. I kept a sample of it in a Kleenex in my purse for awhile, but it smelled so bad, I threw it out," Klementowicz recalls.

Following up on the Concerned Citizens of Greenpoint's early organizing effort, in 2002 Basil Seggos joined Katie Schmid, who worked for councilman David Yassky, in calling together a meeting of local residents concerned about the oil spill to talk about what should be done. Out of this gathering came the Newtown Creek Alliance, a coalition of local residents, environmentalists, and government officials who wanted the oil spill and other pollutants flowing into the creek cleaned up.

At the time, many residents of Greenpoint felt beleaguered by the gentrification of the area, which was making it difficult for their children to afford to stay in the community, and the noxious odors from the oil spill, incinerator, and sewage treatment plant, which were constant irritants that eroded their quality of life. Seggos and Schmid injected new energy into the discussion about what should be done to clean up the area. But a lot of residents who had worked on these issues were overwhelmed, recalls Schmid: "They worked hard all day at their jobs and only had a limited amount of time to devote to looking for ways to clean up the environment. We acted as catalysts and got the discussions going again, but it was the local residents who knew the area best and who made the decisions about how to move forward."

One local resident who attended these early organizing meetings was Teresa Toro, a Greenpoint resident who works on transportation equity issues. "At the first meeting, a lot of people were resigned about the oil pollution in the creek," recalls Toro. "It wasn't so much that we were having trouble getting in touch with our anger about the pollution; we were just stymied." Greenpoint was one of the oldest heavily industrialized areas in the nation, and for years it had been a favored site—some say a dumping ground—for heavily polluting industries and necessary but noxious public works such as incinerators and sewage treatment plants. "It wasn't that we took a Not In My Back Yard (NIMBY) position on all businesses that wanted to move into the area; but we did start to adopt a Not *Always* In My Back Yard (NAIMBY) motto," Toro explains.

One of the first orders of business for the Newtown Creek Alliance and Riverkeeper was to put pressure on local businesses to install modern environmental controls to reduce the amount of toxic wastes flowing into the creek. Most of the small firms were anxious to comply with regulations and just wanted to figure out how they could do it without going broke, Schmid recalls. ExxonMobil was the exception: its position was that it was pumping oil out of the ground as rapidly as possible.

For years residents tried unsuccessfully to push ExxonMobil into a more aggressive cleanup effort. They got little help from state regulators. DEC officials often took the oil company's side at public meetings, residents recall. "The cleanup of the oil spill was not well handled in the past. Regulators should have gone after the oil problem years ago. The Department of Environmental Conservation blindly accepted the reports Exxon made about their cleanup activities instead of pursuing the matter aggressively. Their actions during the early days were really reprehensible. It may sound clichéd to say, but I'm shocked that DEC let this go on as long as it did," she adds.

## Riverkeeper Lawsuit

In January 2004, Riverkeeper filed notice of intent to sue under the Clean Water and the Resources Conservation and Recovery Act. The lawsuit was formally filed on May 15, 2004, against ExxonMobil for polluting Newtown Creek and violating several federal statutes. "After Riverkeeper showed up, everything changed," observes Toro who, with her partner, Rolf Carle, joined the suit as coplaintiffs. When Riverkeeper brought their patrol boat up the creek, media attention began to focus on the issue, and local politicians started getting on board, says Toro.

The goal of Riverkeeper's lawsuit is to require the oil companies responsible for the spill to clean up the mess and for the government to require that it be done in a timely and professional manner, Seggos explains: "We want the oil companies responsible for the spill to commit to an aggressive remediation of the problem, increase the rate of pumping oil out of the ground, remove the oil in the sediment, and prevent more of it from leaking into the creek," he says. Although the lawsuit is not aimed at winning a financial settlement, Seggos wants ExxonMobil to pay for more monitoring wells to be installed and map the full extent of the underground oil plume in order to accurately survey what needs to be cleaned up.[9]

Toro is more explicit about compensation issues: "ExxonMobil owes this community a lot. They have years of atonement they need to do for all the asthma, illness, anxiety, and inconvenience their oil spill caused residents. ExxonMobil is now making charitable donations, such as a $25,000 donation to Greenpoint's YMCA. That's nice but I think it should have been more like $125,000," Toro comments. Asked what she thinks the oil company should be doing in the community, Toro suggests that it start by cleaning up the soil on its own property down by the waterfront and then donate the property to the community for a riverfront park. "There would be some poetic justice if they did that; at least it would be a start" she observes. "Our community stored their oil under our houses for decades. Maybe we should figure out what the going rate is for storing oil, amortize it over the years it was stored here, and then come up with a ballpark figure of what they owe the community."

## Mass Tort Lawsuits

Residents of fenceline communities, such as those in Greenpoint, who suffer from illegal toxic emissions emitted by neighboring factories and military bases, are rarely able to find relief through the courts. Lawsuits brought against polluters for damaging the health and property values of affected residents are hard to win. In many instances it is difficult to find a lawyer willing to bring the case to court. Part of the problem is that these communities tend to be poor and have little political muscle. Furthermore, many contaminated communities fail to attract good tort lawyers because none of them can see a way to make a profit off their cases.

Since it is difficult to prove that a toxic release caused a specific disease, lawyers often try to establish that chemical emissions resulted in damage to property values or quality of life. Unfortunately, property values in low-income neighborhoods are not high enough to make it worth their while to prepare a lawsuit. This appalling legal calculus leaves whole communities facing long-term exposure to elevated levels of harmful chemical pollutants without a good shot at winning legal redress.

By contrast, Greenpoint is one of the few fenceline communities that has attracted toxic tort lawyers. Several mass tort lawsuits were filed by large groups of residents charging ExxonMobil with damage to residential properties and health following the Riverkeeper legal action. One of these, representing the interests of some four hundred Greenpoint residents, was brought by the Los Angeles legal team of Girardi and Keese, made famous in the water pollution lawsuit depicted in the movie *Erin Brockovitch*. Not to be outdone by a team of West Coast lawyers poaching on their turf, personal injury lawyers from the Manhattan-based law firm of Napoli Bern and Ripka signed up clients at Greenpoint's American Legion Hall for a suit demanding $58 billion in damages. One of the parties cited in the suit is ExxonMobil, the most profitable company in the nation.[10]

Industry representatives will likely fight these lawsuits vigorously, arguing that the spill took place at least fifty and in some instances over one hundred years ago when there were few regulations about the disposal of oil wastes. They will point out that the companies that ran the refineries that caused the spills have long since shut down; although they are technically liable for contamination on land they own, they are not morally responsible for it in the same way they would be had they operated facilities in a reckless manner. Industry lawyers will also point out that the clay-bearing composition of soils in the residential neighborhoods in Greenpoint effectively forms a barrier that keeps dangerous vapors from entering residential homes and that no formal health studies prove any health problem in the neighborhood. As for the pollution causing real estate values to fall, they will note that the price of apartments in Greenpoint has appreciated faster than in many other parts of Brooklyn.

## A Case of the Vapors

The oil companies named in the lawsuits are potentially liable for the damage to the health of residents their spillage may have caused. Living atop a lake of petrochemicals could cause disease if toxic vapors make their way up through the soils and migrate into homes through cracks in foundations and concrete slabs. The possibility that elevated levels of benzene, a known carcinogen, may be outgassing into homes is of particular concern. Some indoor air testing has already been carried out, but critics charge that the testing thus far is not comprehensive enough to establish whether indoor air pollution constitutes a hazard in the area.

Riverkeeper's chief investigator, Basil Seggos, asked both regulators and oil company officials if there was methane or benzene in the soil and was initially assured there was not. Seggos did not take them at their word. Instead, he oversaw the drilling of test holes in an industrial area in Greenpoint. A laboratory analysis of the soil boring samples they removed found that they were so heavily contaminated with benzene and methane that they should be classified by regulators as toxic wastes. "The levels of these chemicals we found were one hundred times higher than OSHA short-term exposure standards for industrial workers," says Seggos, referring to the 1,560 ppm of benzene detected in the sample.[11] For his trouble, Seggos was reprimanded by DEC officials and told not to drill any more test holes.[12] Officials at ExxonMobil later paid to have soil testing done in the same industrial area, and the results corroborated Riverkeeper findings and in some cases exceeded them, Seggos says.

However, other ExxonMobil vapor studies concluded that the health of residents was not at risk from petrochemical vapors emitted by the underground oil. An Exxon geologist dug test holes seven to eight feet deep and found that the soils were contaminated in the industrial section of Greenpoint but not in the residential neighborhood. "From the outset of our remediation project, we believed there was no significant health danger to the people of Greenpoint. We don't think that has ever been an issue," Barry Wood, a spokesman for ExxonMobil, told a reporter.[13] Testing by the attorney general's office has also failed to turn up evidence of elevated levels of benzene in fifty-two homes they checked, although they did find traces of solvents in two of the homes. State DEC regulators are currently conducting their own indoor air vapor studies in 250 Greenpoint residences and businesses.

**Drilling Next Door**

Petrochemical vapors became an intense problem for Dorothy Swick when her neighbor had a misguided idea that he could prove he was living over the oil spill by drilling two holes in his backyard a foot away from Swick's property line. The bore holes, made without a permit, were over forty feet deep and punctured the layer of clay that lies over the underground oil.

In July 1996, Swick returned home to find "a terrible odor in my backyard and basement." She called the New York City Department of Environmental Protection and the state DEC. Officials from DEC placed monitoring canisters in her basement and confirmed that petrochemical vapors, including benzene, were accumulating. They subsequently traced the source of the problem to the boreholes next door and filled them.

A crew paid for by ExxonMobil came in and broke holes through Swick's concrete basement walls and installed a system of PVC pipes that collected the vapors and vented them through her roof. To this day, she remains unsatisfied with these remediation efforts. The work damaged her home, and she has never been compensated for it. Nor has the odor problem been entirely solved: "You can still smell it; the petrochemical vapors have a very distinct smell," notes Swick, who keeps an odor journal to document the problem.

Did this exposure compromise her health or that of her late husband? Swick is reluctant to make any categorical comments: "It does burn your eyes, and sometimes you have to put a cloth over your mouth to breathe . . . but I am not a medical person." As for longer-term effects Swick is unsure: "I feel okay now, but who knows about the future," says Swick, who has been a resident of Greenpoint for over seventy years.

In the past, commercial establishments have also wrestled with the problem of petrochemical vapors percolating up from the oil plume. Several warehouses in an industrial area by Newtown Creek had problems with contaminated air vapors, Seggos reports. Owners of these buildings solved the problem by creating positive air pressure in their warehouses, thus keeping the methane- and benzene-laced vapors from seeping into their buildings. If it turns out that there are toxic air vapors beneath some residential buildings, one possibility is to create a network of pipes to suck the vapors out of the basements and keep the homes safe to live in, he suggests. Thus far, however, state indoor air testing has determined that toxic vapors are not collecting in residential properties.

## Health Impact

Is it safe to live on top of millions of gallons of oil with vapors coming up out of the ground? Irene Klementowicz, who started Concerned Citizens of Greenpoint, says she does not know how to answer that question: "I'm no geologist or engineer, so I can't really say if it is safe to live here or not. . . . Many people think it [the oil] is making them sick. You see a lot of your neighbors when you go to the dermatologist. And there are also a lot of women here with breast cancer. But the Department of Health says these medical problems are our fault. They say it is because of all the kielbasa we eat and that we drink and smoke. They say the medical problems are because we are not living right and not because of the pollution," she adds. But Klementowicz does not think the problem is with her or her neighbors' lifestyle or eating habits. "The problem is that they [state regulators] did not enforce the environmental laws that were in place," she charges.

Many residents believe there are too many cancers, including rare bone sarcomas and leukemia, in the area, along with elevated levels of asthma and other health problems. But no comprehensive health study has been carried out. Measuring whether Greenpoint has a public health problem would take at least a health survey and at best a comprehensive epidemiological study that compares disease rates of people living above the toxic plume with comparable populations in an unpolluted area. Because these data are unavailable, residents are left with untargeted morbidity and mortality statistics of limited value and anecdotal evidence.

Since benzene is known to cause cancer, some residents are interested to find out if rates of cancer in their neighborhood are elevated. Available statistics indicate that the overall cancer rates are lower in the Williamsburg-Greenpoint area than elsewhere, but the area encompasses such a large population that identifying a localized problem in Greenpoint, near the most contaminated soils, would likely be obscured.

More specific health statistics show that residents in Greenpoint have the highest incidence of certain specific cancers, including stomach cancer in adults and leukemia in children, a troubling sign since benzene has been proven to cause leukemia. A cluster of rare bone cancers has also been found above the spill, according to some residents. Sebastian Perozzi, who lost his leg to bone cancer, thinks that the fact that several of his neighbors had the same disease is not a coincidence. "It can't be," he says. "There are many cancers that can be caused by many things but bone cancer is different. It's rare."[14]

Faced with a lack of targeted health statistics, some residents have become amateur epidemiologists and are conducting their own health surveys by collecting data about how many of their neighbors have cancer or have died from it. Tom Stagg, for example, identified thirty-six cancer victims on the block where he grew up, twenty-five of whom have died. Among these was his father, who died of pancreatic cancer at age fifty-three. "It's not normal," he told a reporter: "I'm sure it's because of the oil spill."[15]

## A Family Medical History Raises Questions

Laura Hofmann is also convinced that the oil spill created a health threat. A lifelong Greenpoint resident, Hofmann, now forty-nine years old, grew up a block and a half from the former ExxonMobil property. "Our family medical history is like something you'd expect from Area 51,"

Hofmann says referring to a military site where research was allegedly done on unidentified flying objects. Hofmann's mother died of central nervous system lymphoma, a rare brain cancer; her father was diagnosed with progressive supernuclear palsy, a rare brain disease; and their dog had a brain disease as well. In addition, her sister and brother were born with a birth defect known as congenital sphirocytosis. Hofmann's own health and that of her children is not good. "I have lupus [an autoimmune system disorder], and I live in a building where two other residents have already died of the disease," she notes. Her twenty-eight-year-old son also has an autoimmune system disease "that will probably turn out to be lupus," she adds.

"These environmental and health regulatory agencies can stand on their heads until doomsday claiming there is no health problem here, but I will not believe them," asserts Hofmann, who also joined the River-keeper lawsuit as a coplaintiff. When discussing her family's health problems, some of those she spoke with suggest that cancer may run in her family. "But my mother and father weren't related when they got married, and, as far as I know, we're not related to our dog . . . so, for me, that explanation just won't fly."

"All these illnesses in the community are no coincidence," Hofmann speculates. "This area has been a dumping ground for some of the most polluting facilities. If you look back in the records of hearings, you will find that residents have been complaining for years that this area was becoming 'the city's largest toilet bowl' and other comments like that," she says. "They placed these facilities here because it was a low-income area with a lot of immigrants and people here who did not have the political strength to stop it. These were hard-working people who eventually bought their own homes, and many of them were afraid that if they complained, they would lose their homes."

### Inadequate Regulatory Response

"It's really outrageous that the regulators allowed these environmental offenders to offend the way they did," Hofmann says. "They [the regulators] should strong-arm these companies so that they won't pollute any more and so they will clean it up. . . . We have the technology now to go to the moon and the bottom of the sea, and you want to tell me that we can't come up with a technology to clean up the oil in the soil here? No, something better can be done, and it should be up to the oil companies and the regulators to get it done."

For his part, Seggos is careful when it comes to assessing the potential health problems associated with living atop this oil blob. The soil and vapor test data are very preliminary, he notes, and it would be premature to suggest what measures, if any, need to be taken to ensure the safety of the residential population.

Also taking an objective approach to the possibility that the oil spill has caused widespread illness is Christine Holowacz: "I go to a lot of the meetings, and I speak Polish, so I hear a lot of people talk about how their wife is sick with leukemia, how they know several people with brain cancer, and how their basement stinks of oil. But this is all hearsay, and there is no concrete evidence that can prove the oil is causing these problems. There have been no real medical studies, so the health threat is hard to assess," she continues. Furthermore, in addition to the environmental hazard posed by the underground oil, the community was also burdened with numerous other heavily polluting facilities that could have contributed to health problems among residents. Among the larger locally unwanted land uses were the incinerator and a copper smelter. Pollutants from these facilities could also have contributed to the number of cancers in the area, she adds.

Despite these complicating factors, Holowacz thinks that the oil companies should not only be required to suck out the oil from under the homes but should also clean up the soil as well. There are some new technologies. such as using bacteria that eat up the pollutants in the soil, she notes. The money from recovering the free product should also be used to cover the costs of a health survey and to pay for a consultant for the grassroots groups so that they can have someone they trust to monitor the petrochemicals in the soil and vapors.

**Attorney General's Lawsuit**

On February 8, 2007, Andrew Cuomo, the newly elected attorney general of New York State, followed Riverkeeper's lead and submitted notice of intent to sue ExxonMobil Corporation for its pollution of Newtown Creek and the groundwater and soils beneath Greenpoint.

The attorney general's suit alleges that since the early 1900s, Exxon-Mobil has spilled, leaked, or otherwise discarded 17 million gallons of petrochemical product into the area, inflicting "great harm upon the state of New York, its citizens and residents and the environment." Cuomo went on to charge that the spill poses "a significant threat to the health" of residents who use the creek for fishing, crabbing, or recreational activities, as well as those who live nearby. The suit notes that many of the chemicals released, such as benzene, are carcinogenic. Other highly toxic chemicals were also emitted, including touene, xylene, trichloroethylene, vinyl chloride, acetone, and heptane. Furthermore, the suit accuses ExxonMobil of degrading water quality and threatening the welfare of wildlife.[16]

Describing ExxonMobil's Greenpoint spill as "one of the worst environmental disasters in the nation," Cuomo states that the company's "toxic footprint" is found all over the area: "ExxonMobil has proven itself to be less than a model corporate citizen, placing its greed for windfall profits over public safety and the well-being of the environment."[17]

A veritable parade of high-powered New York public officials were quick to add their support for the lawsuit. "ExxonMobil has dragged its feet when it comes to tackling their toxic mess in North Brooklyn," observed New York Senator Charles Schumer, adding that the lawsuit by the attorney general's office would help provide the needed pressure "to compel them to protect the health and well-being of Brooklyn residents." State legislators piled on: "It is high time that the State forces Exxon-Mobil to use some of their record breaking profits to compensate the environment and community that deserves justice," said State Assemblyman Joseph Lentol.[18]

After state officials ignored the heavy contamination of their neighborhood for decades, the attorney general's lawsuit (and the support it received from high-ranking New York state officials) was sweet music to the ears of many Greenpoint residents. "I'm very happy with the aggressive action that the attorney general is taking to clean up the oil and groundwater and soils. This is the first time that someone is actually going to clean up Newtown Creek for the community," says Christine Holowacz, cochair of the Greenpoint Waterfront Association for Parks and Planning.

## Incomplete Cleanup

Prior to the lawsuits, ExxonMobil and Chevron had already begun oil recovery operations and have removed more than 9.3 million gallons of oil, and they continue to operate pump and treat cleanup operations to recover more oil from polluted groundwater. A new technology, vacuum-enhanced recovery, is also under experimental use. However, some critics point out that just sucking out the oil in the groundwater, a process that is likely to take over twenty years, does not remove the source of the problem. In addition to the polluted groundwater, there remains a large volume of oil that clings to sand, rocks, and soil, and this oil will continue to outgas into the homes above and seep into the creek and groundwater for decades to come. Removing these petrochemical-coated soils would require relocating a large number of residents, digging up the neighborhood, and treating or removing the toxins at huge expense. Currently the likelihood of this ever happening seems remote, leaving the prospect that large amounts of spilled oil will remain in the ground for decades to come.[19]

Should residents living over the oil spill be relocated so that soils can be dug up and treated? Most residents are not looking for the oil companies to give them money to move, reports Christine Holowacz. "It's not about money; it's about being able to stay here and know that the grandchildren will be safe living here," says Holowacz, who emigrated from southern Poland to Greenpoint in 1972. "There are many Polish families in Greenpoint, where there are three generations living in the same house. This is not a community where people move out; they like to stay here," she observes.

Seggos is also wary of any talk about relocation: "This is a great New York neighborhood, where residents regularly close off streets and throw block parties." Breaking up such a closely knit community would be a huge loss, he continues. He hopes that a technological solution can be found that will keep the neighborhood intact rather than requiring any relocation.

While it may prove impossible to clean up all the spilled oil, suits filed by the New York attorney general's office and Riverkeeper have hastened efforts to pump out and treat contaminated groundwater that can be reached. In the wake of the lawsuits, petrochemical companies cited as responsible parties are now pumping out some fifty thousand gallons of oil a day, almost three times more than they were required to extract under a settlement reached in 1990.[20]

## Mixed Residential/Industrial Zoning

Looking back, what is clear from the experience of residents in Greenpoint is that permitting heavy industry to be built close to a residential neighborhood was a mistake; at a minimum, it should have created a buffer zone around the industrial area to keep residents at a safe distance. Part of the problem is that for about 140 years, Greenpoint has been a mixed residential/commercial/industrial area, a zoning category that risks exposing residents to dangerous chemicals and causing disproportionately high rates of illness.

Today Greenpoint is poised for yet another change as New York's outward-pushing gentrification brings more middle-income families to the area. To accommodate this influx, old industrial buildings are being torn down and new apartment buildings are going up. There are even plans to build a greenbelt along the shores of Newtown Creek and attract more recreational and marine-oriented businesses to the neighborhood. "With the area gentrifying we now have a whole new generation of families with their kids moving in. How many of these new kids will end up future patients?" worries Greenpoint resident Laura Hofmann.

The pollution problems that residents in Greenpoint and other communities described in this book have experienced point to the need for a more sophisticated land use strategy that separates residential areas from zones where heavy industries handle and emit toxic chemicals. A lasting solution to this problem will require recasting the way we make land use decisions. The current system in the United States, which devolves most zoning decisions to local officials who are frequently beholden to the real estate and development companies that contribute heavily to their campaigns, will need to be modified. Clearer lines between residential areas and industrial zones will have to be delineated, and buffer zones may need to be created. More effective regulation and health monitoring between where people live and where we make our goods will also be required to keep fenceline populations safe.

## Disease Clusters

Far from this East Coast urban setting of Greenpoint, in an isolated desert oasis community in Nevada, residents have faced a mysterious spike in rates of childhood leukemia, raising the question of whether the disproportionately high rate of illness was caused by toxic exposures emanating from a naval air station, an aviation fuel pipeline, high levels of arsenic in the water, heavy metals from a smelter and factory, radioactive pollutants left over from the Cold War, or some combination of these various contaminants.

Disease clusters such as the one described in the next chapter about Fallon, Nevada, are notoriously difficult to solve when efforts are made to trace the cause of the disease outbreak. Government regulators and researchers from the Agency for Toxic Substances and Disease Registry have investigated many such disease clusters but have failed to find clear evidence that exposure to toxic chemical and radioactive pollutants is the culprit. Nevertheless, many other researchers and residents think that common sense and new research provide enough evidence that environmental factors are likely the cause of these diseases that a precautionary approach is warranted.

# V

An Ongoing Puzzle: Disease Clusters Possibly
Caused by Multiple Sources of Pollution

# 12

## Fallon, Nevada: Largest U.S. Pediatric Leukemia Cluster near Naval Air Station and Tungsten Smelter

Vonda Norcutt's family is deep into the rodeo. They rope calves from the saddle, jump off horses to wrestle livestock to the ground, and ride bucking broncos. Norcutt is secretary of the Nevada High School Rodeo Tournament; her husband, Wayne, is a former high school rodeo champion; and both of her children compete.

In the 1980s the Norcutt family lived in Fallon, a small town located sixty miles east of Reno along Highway 50—called "the loneliest road in America." Fallon is a frontier town of seventy-five hundred residents formed during the California Gold Rush when families left Missouri heading west to Sacramento to stake a claim. Among the other dangers they faced on their journey west was a forty-mile stretch of desert badlands just east of Fallon, where deep sands and bad water killed so many mules that in some places, according to one pioneer's diary, you could step from one pack animal's skeleton or carcass to the next without touching the ground.[1]

In this desert oasis, which now hosts casinos, fast food joints, and RV dealerships, Vonda Norcutt worked in the 1980s as a laborer at the Naval Air Station (NAS) at the edge of Fallon where F-16 pilots crack the dry air as they swoop overhead on their way to destroy the rusting hulks of tanks and trucks left as targets on the desert floor. At the time, Norcutt's husband worked for the Nevada Department of Transportation, pouring concrete for the growing network of state roads.

Norcutt, now thirty-nine years old, was in Fallon when she gave birth to their first son, Devon, in 1989. The family subsequently moved deeper into the desert to Cold Springs near Austin, Nevada. There, at the age of seven, Devon became noticeably lethargic, and Norcutt took him to the doctor, where blood tests were done. Told that her son was healthy, she decided to seek a second opinion. Her doctor in Fallon did further blood work and advised her to pack her bags and take Devon to the hospital in Oakland, California, for more tests to see if he had leukemia. Doctors in Oakland diagnosed Devon as having acute lymphoblastic leukemia (ALL) in 1997. "They said it was not a death sentence, but it was a leukemia cancer nonetheless," Norcutt recalls.

"The first year of therapy was hell," Norcutt remembers. Devon endured months of chemical and radiation therapy. Norcutt had to give him injections in his thigh, and Devon's hair fell out as a result of the treatment. "He was very brave about the injections, but it was hard on him," she recalled. "He did not know what leukemia meant and only became really scared later when it learned it was a cancer."

Learning of Devon's leukemia, Navy pilots from all over the United States showered him with letters containing photos of them standing by their fighter jets. They also presented him with a specially tailored flight suit, gave him a personal tour of the Fallon's "Top Gun" base, and published an article about him in a Navy magazine. Devon recovered and now, at seventeen years old, competes in rodeo events. Not all the children in the Fallon leukemia cluster were so fortunate: three died.

It was not until the Norcutt family moved back to Fallon in 2000, where they found jobs at the power company, that Vonda Norcutt began to suspect that her son's leukemia might have been caused by environmental contamination. "At one point in 2000 and 2001 there were kids being diagnosed with leukemia every six weeks," Norcutt says. "We knew something wasn't right."

## Health Impact

Norcutt was not wrong. Joseph Wiemels, a genetics researcher at the University of California at San Francisco (UCSF), describes the Fallon pediatric leukemia cluster as the most striking cluster ever studied.[2] The average number of leukemia cases per 100,000 children ages birth to nineteen is 4.3 per 100,000, and some 3,000 are diagnosed in the nation every year. The projected incidence of childhood leukemia in a town the size of Fallon is one every three years.[3] Instead, Fallon played host to at least 17 cases over the seven years from 1997 and 2004. Others believe the Fallon leukemia cluster is larger. Mark Witten, a biomedical research professor in the Department of Pediatrics at the University of Arizona's College of Medicine, counts 20 cases in Fallon.[4] Families in Search of Truth (FIST), a group of Fallon families whose children had leukemia, calculate that there have been 25 cases and 5 deaths.[5]

What explains these different counts? One issue is how long a family stayed in Fallon and whether to count families who moved out. Another controversy surrounds whether to count the case of an undocumented resident. There is also the question as to when to stop counting. Some residents claim the cluster is not over and point to the case of Rebecca Rau, a fifteen-year-old girl recently diagnosed with leukemia. Even using the lower official number of seventeen cases, there are eight times the number of statistically expected leukemia cases in a town that counted 7,537 residents, including 2,383 children, in 2000. The odds are 232 million to 1 that chance could explain the size of this cluster calculate the authors of a peer-reviewed article published in *Environmental Health Perspectives*.[6]

### Eric Heavens

The case of Eric Heavens, who visited Fallon only in the summers, illustrates the subtlety of the question as to who should or should not be included in the leukemia cluster. Heavens has not been counted as part of the cluster despite clear evidence that he had ample opportunity to be exposed to a number of the town's contaminants.

Eric was diagnosed with ALL leukemia when he was fourteen and died of the disease a year and a half later in May 2000. His father, Bill Heavens, a retired wholesale florist from Sacramento, took his son to Fallon every summer to stay with his aunt and uncle who had a place in the tiny settlement of Stillwater, near the Naval Air Station, sixteen miles outside Fallon. After one of a number of summer visits, Eric became tired and skinny before his father took him to a doctor in Sacramento where they lived. "He had a lot of benzene and heavy metals in his system, while his sister, who never went on the trips to Fallon, did not," observes Heavens, who thinks it is obvious that his son was contaminated in Fallon. The boy was very sick by the time he was diagnosed and was immediately given transfusions, chemotherapy, and radiation treatments.

The treatments damaged his heart before his kidneys shut down. "I had to just stand there and watch him die. It got really ugly," Heavens says.

After his son was diagnosed with leukemia, Heavens began to attend some of the meetings in Fallon at which officials from the Centers for Disease Control (CDC) and the state Department of Health talked with concerned citizens. "I stood up to some of the CDC people" who tried to minimize the problem, he recalls. Officials were looking for a single cause "when in fact it was the combined impact of a variety of contaminants including benzene, tungsten, arsenic, and radioactive pollutants which formed a 'witches' brew' that was killing kids," he adds.

### Dustin Gross

On April 17, 1999, Brenda Gross's son, Dustin, who was born in Fallon in 1995, was the second child in town diagnosed with ALL leukemia. Her son, now eleven years old, plunged into multiple injections of chemotherapy drugs and spinal taps. "Dusty" Gross had a device installed to facilitate the delivery of medicines into his body and the family learned how to flush it out regularly to avoid infection. "Dustin's leukemia brought the family closer together," observes Brenda Gross, who would just as soon that drawing closer had happened for a different reason.

When Gross, a water rights specialist for an engineering firm in Fallon, heard of a third leukemia case in town, she contacted Barbara DeBraga, the oncology nurse at the local Banner Churchill Hospital. She asked DeBraga if it did not seem strange that three childhood leukemia cases had been diagnosed in such a small town. Cancer clusters such as this sometimes occur randomly, DeBraga told her, but she promised to bring the cases to the attention of the Nevada State Health Division. She also informed her sister, Marcia DeBraga, an assemblywoman in the state legislature, who subsequently pushed for an investigation.

As the size of the leukemia cluster in Fallon became known in late 2000 and early 2001, a number of regulatory groups dispatched teams of experts to town, including the Nevada State Health Division and the Nevada Department of Environmental Protection. Researchers were careful to keep expectations among residents low. Federal CDC officials from the Agency for Toxic Substances and Disease Registry (ATSDR) told the parents of leukemia-diagnosed children that they would carry out research, but that it would take a long time and that residents might never know what had caused the leukemia cluster. "And that is exactly what happened," Gross notes.

ATSDR investigators conducted surveys with more than five hundred variables among 205 residents in 69 Fallon families in March 2001. Blood, urine, and cheek samples were taken and tested for the presence of 139 chemicals. Air, water, and dust samples were collected from 80 homes and analyzed for the presence of some 200 chemicals. The results showed elevated levels of tungsten and arsenic in both urine and water samples. In fact, the levels of tungsten in the urine of children diagnosed with leukemia in Fallon was, on average, eighteen times that of other Americans and twice as high as Fallon adults.[7] Nevertheless, regulators found "no exposure consistent with leukemia risk."[8]

"It was pretty discouraging," says Gross, who volunteers in Fallon on the Farmer's Market Board and helps arrange the cantaloupe festival every year. The CDC no longer does research on cancer clusters because they had failed in the past to find out what caused them and funding dried up, she explains. Furthermore some of the techniques used to test for contamination seemed to lack common sense, she observes. Technicians took dust samples, but they often captured them in places where she vacuumed and dusted regularly. And when they tested her tap water, technicians let the water in the sink run for twenty minutes before sampling it. "I don't let the water run that long before I take a drink," Gross notes. Even the sample of water they took out of her well seemed odd. Instead of opening the well and scooping a sample off the surface to see if it had benzene or other petrochemical products, they ran the pump and churned up the water before testing. "I understand the need for protocols, but this didn't make a lot of sense," Gross observes.

## Adam Jernee

In 1999 Richard Jernee, recently divorced, relocated to Fallon with his nine-year-old son Adam. Jernee moved into a subsidized unit at the Sunridge Quarters apartments in Fallon and found a job driving a water truck for A&K Excavating Company. Life was getting better, Jernee recalls: his son's grades were improving, and he was just days away from putting money down on a new trailer so they could move out of public housing into their own home. Then his son was diagnosed with ALL leukemia. By that time, they had been in Fallon for a year and a half.

Adam's health problems began as a chest cold that just would not quit. Doctors would give him antibiotics, and he would get better briefly, but then the coughing would start again. Finally Jernee took him to a doctor and told him that the problem was recurring and that his son should be thoroughly checked. Jernee was stunned when doctors told him his son had leukemia. "At that point the doctors just took over. They told us we were going to Oakland [the hospital in Oakland, California]. If I'd had medical insurance and better finances perhaps I would have had something to say about it," he muses.

At the Oakland hospital, Jernee was reassured when doctors told him that there was a good chance they could give Adam medicines that would make him better. "I figured he would make it through, and I would get my son back," he remembers. During one of his son's treatments in Oakland, doctors told Jernee that they were treating another Fallon resident, Sarenyah Rivers, daughter of Carinsa Phelan, who lived in the same apartment complex with the Jernees. That was the first time Jernee heard that his son might be part of a leukemia cluster. "I had no idea at the time what a leukemia cluster was," he says.

After multiple treatments Jernee's son began to fail. His lungs, scarred from his recurrent respiratory illness, could not provide him with the air he needed, a problem exacerbated by radiation treatment. He was placed first on steroids and then on a respirator. Finally doctors told him that there was no hope for his son. When Adam regained consciousness briefly and Jernee told him that he loved him, his son acknowledged that he had heard his father with a nod. The next day the breathing machine was turned off. "He opened his eyes once after that. It was very distressing and very sad," Jernee says. "You never think that it will happen to you and your children," he told documentary filmmaker Amie Williams, "but when it does nothing in the world can make it better."[9]

After Adam Jernee's death, residents in town became increasingly polarized over their leukemia cluster problem that was now being reported on nationally televised programs. At first residents in Fallon seemed to care about what he and his son were going through, recalls Richard Jernee. However, when news of the leukemia cluster hit the front pages of newspapers, property values in Fallon started to slip. Jernee was quoted in a number of articles, and some of his neighbors thought he was giving Fallon a bad name and hurting business and real estate values. One resident warned him, "You are not going to be around here for long if you keep talking like that," Jernee says. Life in town became uncomfortable: "Friends dropped me, and people who had offered to help suddenly did a 180-degree turn. Then someone poured motor oil on my porch and my car. I left because I was depressed and suicidal. I left to start a new life. It was the right thing for me," says Jernee who has since moved to Washington State, where he works in an automotive repair shop.

"I'm angry, but I'm not mad at the people of Fallon. I feel bad for them," says Jernee. "Fallon is a robust little town with lots of jobs. What is the federal government going to do? Shut it down?" Jernee thinks not. People just turn their heads: they do not want to face the problem, he says. Jernee faults Kinder Morgan Energy Partners, the jet fuel pipeline company, for not maintaining its pipeline properly and causing spills, and Kennametal, the owners of the tungsten smelter and manufacturing plant in Fallon, for spewing toxic wastes into the air. He thinks someone should pay for these problems and joined with Floyd Sands, another father who lost a child in the Fallon leukemia cluster, in filing a wrongful death lawsuit against Kinder Morgan Energy Partners and Kennametal. Officials from both firms have denied any culpability in contributing to the Fallon leukemia cluster.

### Sarenyah Rivers

Carinsa Phelan also lived in the public housing complex just a few doors from Adam and Richard Jernee. Phelan was twenty-three when her daughter, Sarenyah, was born in February 1997. Two and a half years later she read *Chicken Soup for the Mother's Soul,* a book that tells how First Lady Barbara Bush first noticed that her three-year-old daughter, Robin, had leukemia. When Phelan's own daughter, Sarenyah, fell sick with a cold, became pale, began to stare off into space, bruise easily, experience frequent nosebleeds, and ask for naps that she had never wanted before, Phelan recognized the symptoms. "I think my daughter has leukemia," she told her doctor.

She was right. On March 30, 2000, her daughter's diagnosis was confirmed, and Sarenyah was immediately started on a transfusion to stabilize her because her blood counts were so bad. "My world fell apart. I was four months pregnant with my son. I just lost it and went into a kind of shock," Phelan recalls. Then they were whisked off to the hospital in Oakland for chemotherapy treatments. Phelan tried to prepare her three year old for the effects of the treatments telling her she would lose her hair but that it would grow back. When the time came, "we just sat there and pulled her hair out," Phelan remembers. Her daughter's teeth became discolored due to the therapy, and she was subsequently diagnosed with learning disabilities, which Phelan thinks may have been caused by the chemotherapy.

Phelan says she was treated well by people in town and that financial help from both the state of Nevada and the local charity, Fallon Families First, was critical in helping her get through the treatments for her daughter. "If it wasn't for them, I don't know what would have happened," she says. Unlike Jernee's experience, no one told Phalen to be quiet when she started giving interviews and talking about the possible causes of her daughter's leukemia. "If anyone had, I think I would have talked even more," she says.

Despite the kindness of local residents, Phelan thinks there was a cover-up about the contamination. She witnessed Kinder Morgan employees working at night under portable klieg lights to patch the jet fuel pipeline. "I asked why they did it at night and was told maybe it was because it wasn't so hot at night. But other construction work was done during the day," she points out. She also faults Kennametal for its massive releases of tungsten, evidence of which is so extensive that it can be seen in satellite photos, and the EPA for failing to adequately regulate the company and ensure that it operated in a safe fashion. Although she is not part of any lawsuits, Phelan believes that Kennametal and Kinder Morgan should pay for the damage she believes they caused. "There is contamination, and children are getting sick. I think these people are keeping two sets of books: one for the regulators and one for themselves," she opines.

By this point, the number of children falling ill with leukemia was attracting national attention. Bill Moyers featured Fallon and Sarenyah Rivers on an edition of his show *NOW* in which he posed the question: "Are we poisoning our children?" The increase in the incidence of childhood cancers over the past twenty-five years "signals that something is going wrong," Dr. Philip Landrigan told Moyers. Landrigan is the director of the Center for Childhood Health and the Environment at Mount Sinai School of Medicine. "I see the cluster of cases of childhood leukemia in Fallon as part of the broad increase in various forms of childhood cancer in the United States, leukemia among them," continues Landrigan. Contaminants are becoming an increasingly important source of disease among children, he observed. Sixty years ago, most illness among children was caused by infectious disease, but today pediatric illness is more likely to be chronic, he added. Pound for pound, the exposure of children to toxic chemicals and heavy metals is greater than that of adults because they eat, drink, and breathe proportionately more than adults; because they are more likely to come into intimate contact with contaminants; and because they are closer to the ground, touch everything, and put their fingers in their mouths. If children are exposed to toxins at critical moments during their development, the result, he says, can be "toxic effects that are absolutely unique and absolutely devastating," he adds.[10]

## Anastasia Warneke

No one knows the Fallon leukemia cluster story in greater detail than Frank X. Mullen Jr., a reporter for the *Reno Gazette-Journal*, who wrote 150 dispatches on the subject. In the course of this reporting, Mullen became convinced that something was causing children in Fallon to get sick, and he kept their photos next to his desk to remind him why he was pursuing the story with such assiduity.

"We don't know how many Fallons are out there," Mullen observes. He looks at his work as an attempt to track a serial killer. There have been efforts to play down the story and mitigate the negative impact on the town by referring to the health problem as a Churchill County, and not a Fallon, leukemia cluster, as well as an effort to sidestep the question of whether the jet fuel pipeline that runs through Fallon might have leaked and caused illness, he notes.

One of Mullen's articles focused on Matthew Warneke and his daughter, Anastasia, who was eight years-old when she was diagnosed with leukemia and subsequently successfully treated for the disease. Anastasia's case is of particular interest because she lived in Sierra Vista, Arizona, where she began to exhibit symptoms of leukemia before moving to

Fallon. Two weeks after she arrived in Fallon, she was diagnosed with the disease. For bureaucratic reasons, which do not make a lot of sense, her case is counted in the Fallon cluster and not among the cases attributed to Sierra Vista. Her father thinks this is ridiculous, noting that it is highly unlikely that she contracted leukemia after only two weeks in town.

Mullen's article about Matt Warneke struck a nerve with Amie Williams, a documentary filmmaker who was living at the time in Reno. The article depicted a blue-collar guy who climbed power poles to make repairs for his living and his daughter, who was diagnosed with leukemia, Williams recalls. "There was something about the photo that got to me," she says. Mullen's article launched her on a project directing a multiyear documentary entitled *Fallon, NV* that followed the Gross, Jernee, and Warneke families as their children were injected with chemicals, anesthetized for spinal tap procedures, and treated for infections.

Williams's intimate look at the human cost of leukemia is at times hard to watch. For example, there is the moment when young Anna Warneke, who is playing outside, suddenly runs away from the camera, turns her back, holds onto a tree, and vomits from the nausea caused by her chemotherapy. Her father comes to her, holds her, and tries to explain what is happening.

**Stephanie Sands**

One of the most agonizing stories in the Fallon leukemia cluster is that of Stephanie Sands, who lived in Fallon for several years before moving with her father, Floyd Sands, to Pennsylvania. Her son, Ewan, was one year old when Stephanie, then nineteen, was diagnosed with ALL leukemia of the T-cells, a particularly deadly type of leukemia. This same extremely rare variety of leukemia was the diagnosis for four children in the Fallon leukemia cluster, including three who died.

In January 2001, while surfing the Internet, Floyd Sands discovered that his daughter's leukemia was part of the Fallon cluster. Three months later, following an umbilical cord blood (UCB) stem cell transplant, Stephanie Sands's cancer metastasized to her brain and spinal cord, and she died on September 1, 2001. She had suffered for two years from the disease, including bouts with internal bleeding, septic shock, and infection.

After his daughter's death, Floyd Sands devoted much of his time to researching the unanswered questions about contamination in Fallon and its possible link to the leukemia cluster. Sands believes that there is a conspiracy to hide the truth about the Fallon cluster. He wants to find out what killed his daughter and sickened the other children in Fallon, and he suspects powerful forces are aligned against him and other residents seeking this truth. He is convinced that contamination leaking from the pipeline that transports jet fuel to the base and tungsten dust from the Kennametal plant played a role in causing the leukemia that killed his daughter.

The pain of bereavement does not necessarily make one objective. Clearly just because Sands and the other parents of children stricken in the Fallon cancer cluster are agonized and angry does not give them credentials to solve a subtle public health mystery that has eluded medical experts. However, in Sands's case, the pain of losing his daughter did engender persistence in doggedly pursuing an answer. The CDC has a 100 percent failure rate in looking into cancer clusters, having failed 109 times consecutively, Sands observes. Let them pursue their investigations, he continues, while he follows his own path. "Even a blind squirrel will find a nut occasionally," he observed to a reporter from Las Vegas.[11]

Sands disagrees with the official calculation of seventeen children in the Fallon cancer cluster. He claims no fewer than twenty-five children from Fallon have been diagnosed with leukemia, and five have died. "This is the most aggressive attack of cancer in medical history worldwide in terms of time and spatial clustering," he argues. Local officials downplay the problems and intimidate the residents into silence, Sands charges. Public health officials in Fallon are into "the three D's: deny, discredit, and discount," when it comes to health problems caused by toxics, he observes.

In October 2002, Sands joined FIST, Fallon's leukemia cluster grassroots group, which was organized with the goal of supporting affected families and uncovering what was causing this disease that was sickening and taking the lives of children in their community. He currently runs a Web site, Fallon Cancer Crisis, which provides residents and reporters with a chronology of events in Fallon, dates when children were diagnosed or died from leukemia, and up-to-date information on regulatory activity, scientific studies, and lawsuits.

## Inadequate Regulatory Response

A comprehensive chronology of the Fallon leukemia cluster, published by Reno-based reporter Frank Mullen Jr., gives an overview of how events unfolded. The story begins in April 2000 when a Nevada State assembly-woman, Marcia DeBraga, having learned that four cases of leukemia had been diagnosed in Fallon, requested state officials to initiate an investigation. "This is the sort of thing that makes the hair on the back of your neck stand up," Randal Todd, the Nevada State Health Division epidemiologist who ran the investigation, told a Tucson reporter.[12]

In June 2000, state health officials announced an active investigation into the case and began to prepare a thirty-two-page questionnaire designed to see what the families of children who came down with leukemia had in common. Families were asked what cleaning products were used in the home, whether alcohol was consumed during pregnancy, what schools were attended by children, and about their medical history, among a host of other queries. The interviews took some three hours to conduct. The results of this strenuous effort were disappointing. The only commonality found among the families was that they all live in Fallon—something everyone already knew.

One of the only public officials pushing aggressively to get to the bottom of the leukemia cluster was DeBraga, who in mid-February 2001 held a legislative hearing at which state lawmakers called for environmental testing in Fallon. Unfortunately, DeBraga's bill, requesting $1 million in state funds to pay for a study, never passed. Then a month later, state officials issued an "expert panel" report conceding that the cluster was probably not random and that the cause could be a virus or environmental contaminants.[13] Senator Harry Reid (D-Nevada) and Senator Hillary Clinton (D-New York) held a committee hearing in Fallon, following which Reid called on the CDC to investigate.

## No Cause Found

By February 2003, state and federal officials held a town meeting to announce that their investigative efforts had failed to find a cause for the Fallon leukemia cluster. No one was surprised. While their eighteen-month leukemia cluster investigation was the most intensive ever undertaken, the expert panel found no links between the high rates of childhood leu-kemia and the levels of tungsten and arsenic in Fallon's air and water, leaks from the jet fuel pipeline, pesticide spraying, or prior mining and nuclear testing.[14] They did, however, recommend that the town's drink-ing water be treated to reduce its arsenic content and for residents to drink bottled water.

Several of the parents of children in the leukemia cluster were blunt in their criticism of the federal and state investigative effort. "We still haven't seen anything like a competent investigation. . . . So far no one has had either the skill or the will to really look for answers," Floyd Sands told a reporter. Matt Warneke agreed, noting that residents had been warned not to expect investigators to find the cause of the disease cluster; the negative results had been "a self-fulfilling prophesy" and the whole effort a farce, he charged.

The failure of CDC to find a cause of this leukemia outbreak is hardly surprising given its "long and depressing history with cancer clusters," notes Tucson journalist Renee Downing. "Between 1961 and 1982, the CDC investigated 108 cancer clusters; not one investigation determined a cause," she writes. In 1990, after three decades of failure, ATSDR of-ficials recommended that no more clusters be investigated.

Critics of the government investigation find fault with the process on a number of counts. To begin, it was not government health or regula-tory officials who discovered that the cluster existed, notes Floyd Sands, the father of one of the Fallon leukemia victims. Instead, the whistle was blown only after Fallon parents recognized each other at the hospital where their children were being treated.

Government officials also responded too slowly, they contend. Offi-cials should have immediately conducted an environmental inventory of possible sources of contamination in Fallon and taken blood and urine samples from the children diagnosed with leukemia, as well as their par-ents and siblings. Unfortunately, these blood and urine samples were not collected from the children who were diagnosed with leukemia until after they had begun chemical and radiation therapies that changed their body chemistry. Similarly, testing was not done on their homes to see what pollutants they might contain until after many of the houses had already been scrubbed cleaned to protect the children, whose immune systems had been compromised by radiation and chemotherapies, from infection.

## Why Disease Cluster Investigations Are Tricky

Was the failure of the CDC to find a cause due to incompetence? Does the ATSDR intentionally come to "inconclusive" conclusions about disease clusters, as some critics of the agency have charged in a paper titled "Inconclusive by Design"?[15] Or are most cancer clusters inherently unsolvable, as some experts suggest?

To the layperson, it seems obvious that if a number of cases of a rare disease are concentrated in one place, something must be wrong. But that is not necessarily the case, for a number of reasons. First, most disease clusters are due to chance in the same way that pool balls cluster on the table after the break. It would not be unusual, for example, for one area to experience three times the average cancer rate while other areas experience below-average rates. Over time, statisticians assure us, these disparities average out.

But just because some clustering of disease can be expected does not mean there are not other instances in which further inquiry is warranted. In some clusters, the time and spatial concentration of illness so far exceeds what one would predict that the situation calls for an investigation. Such was determined to be the case in Fallon. Furthermore, some major breakthroughs in the treatment of disease have resulted from the painstaking study of disease clusters, particularly those caused by an infectious agent. For example, the mapping of homes with cholera cases in London in 1854 led to a breakthrough in the prevention of that disease. More recently, in the 1980s, a cluster analysis led to the identification of HIV, the virus that causes AIDS.

Some experts believe that the study of cancer clusters may eventually lead to similar breakthroughs. "Since this is a rare cluster [in Fallon] it may have a rare cause," suggests genetics researcher Joseph Wiemels. Even Randal Todd, the state epidemiologist who studied Fallon, said that although he was pessimistic that the cause of the cluster in Fallon will ever be found, he thought it warranted investigation because of its disproportionately large size.[16]

### Elevated Exposure

With government investigators uncovering no smoking gun pointing at what caused the leukemia cluster in Fallon, some of the parents of affected children began to despair that a cause of the cluster might never be found. It was at this point that a number of Fallon activists learned of the work of Mark Witten, a toxicologist and professor of pediatrics at the University of Arizona, who had conducted experiments on the pre- natal effects of tungsten and arsenic on mice.

For many years tungsten was thought to be biologically inert and thus did not represent a health threat. More recent studies, however, dem- onstrate that tungsten does in fact have the potential to cause illness. Alexandra Miller at the Armed Forces Radiobiology Research Institute demonstrated that "tungsten metal alloys are genotoxic to human cells in culture."[17] The military is interested to see if tungsten is dangerous because tungsten alloys are used in artillery shells. A military study conducted in 2005 found that tungsten implanted in animals caused rhabdomyosarcoma, a fast-growing cancer of the soft tissue.[18] The Na- tional Toxicology Program is also testing the carcinogenicity of tungsten through animal studies, and the combination of cobalt and tungsten car- bide is now listed as a probable carcinogen by the International Agency for Research on Cancer.[19]

Witten thinks there is strong evidence suggesting that leukemia can be caused by prenatal exposure of the fetus to contaminants such as tung- sten, cobalt, and arsenic. His hypothesis is that "in-utero exposure to tungsten and arsenic promotes long-term silencing of the tumor suppress- ing genes that are specifically related to B-cell leukemia."[20] Furthermore, studies had shown that human cells exposed to tungsten and a combina- tion of tungsten and cobalt caused mutations.[21] This led to the hypothesis that some combination of contaminants, such as tungsten and cobalt (or tungsten and the benzene in jet fuel), could explain the childhood leuke- mia cases in Fallon.

The ATSDR looked into the tungsten hypothesis and concluded that the heavy metal "is not a concern for cancer." The researchers based their finding on a study of three Nevada towns other than Fallon with high tungsten levels. None had leukemia clusters, they pointed out. But Witten remains unconvinced by the CDC's reasoning and thinks that tungsten may have been a cofactor in causing the leukemia. The ATSDR study found tungsten in the bodies of children in Yerington, Nevada, that were eight times the average level. By contrast, children in Fallon had eighteen times the normal level, he points out.[22] Thus, children in Fallon were getting a much higher dose of tungsten than those in other towns, and the higher level of exposure may have been a cofactor in making them ill.

There was another reason to suspect that tungsten might have played a role. Tungsten levels were found to be elevated in three other communities with leukemia clusters: Sierra Vista, Arizona; Elk Grove, California; and Hoisington, Kansas. "We know for sure that the trees are taking up more tungsten in the last five years than they did 20 years ago. And there's been more leukemia in the last five years. What are the odds of that happening by chance in four different places?" Witten asked a reporter.[23]

Witten and other researchers were struck by the commonalities between Fallon and Sierra Vista, Arizona, both old mining towns with high levels of tungsten, mercury, and arsenic.[24] The childhood leukemia cluster in Sierra Vista, which began in 1997, officially includes seven cases of childhood leukemia and one death. As in Fallon, the number of leukemia cases in Sierra Vista is a matter of contention, with some estimates reaching as high as twelve cases in this town of forty thousand located seventy miles southeast of Tucson. (Arizona does not count cases of children above the age of fourteen or those who had moved out of town.) If there are in fact twelve cases that should be attributed to Sierra Vista, that would be eight more than would be expected statistically, notes reporter Renee Downing.[25] Also striking is that fact that Sierra Vista, like Fallon, is a military base town, playing host to the Fort Huachuca military air field, and both towns have pipelines passing through them that carry JP-8 fuel.

## Sources of Tungsten in Fallon

The presence of tungsten in Fallon is an established fact. Fallon is in tungsten country, and it is home to inactive tungsten mines as well as an active smelter and manufacturing plant. Ten miles north of Fallon, the impact on the land of Kennametal's tungsten smelting plant on the desert is hard to miss. Granted a waiver by state officials, Kennametal operated an open-air industrial kiln that refined tungsten ore into tungsten carbide, without modern environmental controls, from 1969 to 1994. As a result, the hills and desert surrounding the plant are blackened from by-products released from the plant's stacks. In 1994 the EPA put an end to open kiln smelting and required plant operators to upgrade their pollution control equipment. After installing the new equipment, at a cost of $1 million, the plant's emissions were reduced by 98 percent. In 1995 the EPA fined the company $938,000 for violating air pollution standards; the company eventually paid $425,000 in a settlement with the agency.[26]

Kennametal also has a manufacturing plant in Fallon that some researchers think may be a point source for the tungsten exposure in town. The Kennametal plant, located near the Northside School, manufactures tungsten carbide products. Air samples taken at six locations in Fallon in 2004 by Mark Whitten and Paul Sheppard found elevated levels of tungsten and cobalt. The amount of tungsten and cobalt in the air decreased the farther away one traveled from the plant. The researchers also concluded that the tungsten and cobalt in Fallon's air was from anthropogenic, and not natural, sources.[27] Other studies by Witten and Sheppard, using eleven air sampling units, revealed thirteen times the amount of tungsten in the dust in Fallon than in three other towns and one city in Nevada.[28]

It is most likely not a coincidence that in 2004, following the release of the report described above, Kennametal's owners decided to installed high-efficiency particulate air (HEPA)filters. This pollution control equipment reduced the amount of metallic dust that escapes the manufacturing plant to zero, a Kennametal spokesperson claims. "We at Kennametal would like to see an answer [to the cluster] as much as anyone else," Gary Peterson, manager of the plant, told a reporter.[29]

No one knows how much tungsten was emitted by Kennametal's smelter during the two decades that it operated without environmental controls because no monitoring equipment was in place. But there seems little doubt that large volumes of tungsten were released in contaminated water and air pollution from the plant. What impact this had on the health of local residents no one knows for sure. But it is of interest that testing of Fallon's groundwater, tap water, and air revealed tungsten levels elevated above those of other communities. Furthermore, 80 percent of residents tested in Fallon have elevated levels of tungsten in their urine at a rate that is at least tenfold the national average.

Despite these indications that the town was heavily polluted with manmade tungsten and cobalt, regulators and health experts involved in the official inquiry into the Fallon leukemia cluster never directly considered whether the plant's operations might have caused the leukemia cluster. The rationale given for this apparent oversight is that the plant had installed its pollution control equipment in 1994, three years before the first case of childhood leukemia in town attributed to the cluster in 1997. But that still left the decades of tungsten and cobalt pollution that the plant had let loose on the local environment. In retrospect, the decision of the expert panel not to look into the possibility that tungsten from the Kennametal plants might have played a role in the leukemia cluster seems a glaring oversight.

## Intoxicated Trees

While Sheppard and Witten had proven that the air in Fallon was currently heavily burdened with tungsten and cobalt, they had not demonstrated that polluted conditions existed just before the outbreak of leukemia. To test their theory that some combination of tungsten and cobalt might have played a role in triggering the leukemia cluster in Fallon, the researchers had to find out what had been in Fallon's air in the past.

Paul Sheppard, who works at the University of Arizona's Laboratory of Tree Ring Research, was the right man for this job. An expert in dendrochemistry, the study of element concentrations through time in tree rings, Sheppard could tell what levels of tungsten had been in the air in past years by studying the amount trapped in the tree rings. In Fallon, Sheppard took core samples of cottonwood trees located near the Kennametal factory, which uses tungsten carbide and cobalt to harden steel, as well as from trees adjacent to schools. He took special precautions to ensure that his samples were not inadvertently contaminated by boring tools made out of tungsten-hardened steel. To this end, he peeled the tree core samples with tools that contained no tungsten, including a laser trimmer and a ceramic knife.

Sheppard found that tungsten levels in Fallon "quadrupled between 1990 and 2002 whereas the amount in tree rings from nearby towns remained the same."[30] This sharp increase in already elevated levels of tungsten in Fallon's air occurred prior to and during the beginning of the outbreak of the town's leukemia cluster. The elevated tungsten levels were found in a 1.8-mile radius that included many of the town's residences and schools. A further analysis of the tungsten particles determined that they were man-made and not naturally occurring and that they were small enough to be inhaled and potentially cause health problems.[31] Cobalt levels were also elevated during this period, he adds. In addition, Witten and Sheppard found traces of chemicals that could have been deposited by jet fuel in the tree rings.[32]

Following the release of this information, Fallon officials decided to cut down and haul away all the trees located adjacent to four different schools in town that provided the evidence of past elevated levels of tungsten and jet fuel in the town's air. Were city officials disposing of the evidence, some residents asked? Whatever the reason, the University of Arizona researchers already had their core samples.

Sheppard and Witten are careful in interpreting their findings. The fact that tungsten and cobalt were elevated in Fallon's air just prior to the start of the leukemia cluster in 1997 is intriguing "but by itself does not establish a cause and effect linkage between tungsten and leukemia," Witten notes. However, when the tree ring information is added to the fact that elevated levels of tungsten were found in the urine of town residents, this information elevates tungsten to a contaminant of concern in the investigation. More research was needed before a definitive causal link could be proven.

## Perinatal Damage to DNA

Witten and Sheppard were not the only research scientists intrigued by the leukemia cluster in Fallon. Jill James, a professor in the Department of Pediatric Medicine at Arkansas Children's Hospital Research Institute in Little Rock, wrote a paper suggesting that a number of environmental contaminants may have acted synergistically to cause leukemia in children whose immune system had been weakened while in utero or during early development.[33]

Tests by the CDC had demonstrated that arsenic, tungsten, cobalt, cadmium, uranium and mercury, and JP-8 fuel were elevated in the blood and urine of Fallon residents. Other studies found arsenic, antinomy, tungsten, cobalt, and uranium in elevated levels in the soil and water in town and in the blood and urine of residents, James notes.

These contaminants may have damaged the DNA of Fallon children during the "uniquely sensitive perinatal period," she writes. Oxidative stress may have induced the DNA damage that led to the leukemia, she continues. All six cases that she studied of children in Fallon with leukemia "exhibited metabolic biomarkers of chronic oxidative stress." Oxidative stress, occurs when the generation of oxygen-free radicals, which can damage tissues, exceeds the body's antioxidant defense capacity, she explains.[34] Exposure to heavy metals such as arsenic, cobalt, uranium, lead, and mercury can increase the number of free radicals causing oxidative stress, she adds. Exposure to JP-8 fuel can cause the same problem.

Summing up her hypothesis, James argues that simultaneous and multiple exposures to sources of oxidative stress (such as heavy metals in the drinking water, soil, and air) "could interactively increase vulnerability to DNA and chromosomal damage, thus increasing risk of leukemia in genetically susceptible individuals." Although James's research does not definitively solve the question of what caused the leukemia cluster in Fallon, like the work of Witten and Sheppard it offers promising avenues for further research.

James's work is complemented by studies being directed by Frederica Perera at the Columbia Center for Children's Environmental Health. Perera demonstrated how benzopyrene, a chemical formed by burning fuels, can link onto a child's DNA, forming an adduct. The greater the number of adducts, the greater the risk of cancer, he says. If adducts are not repaired properly, a mutation can occur in the DNA, which changes the coding sequence, resulting in deformities or diseases such as leukemia or other cancers. This could be the mechanism at work creating leukemia in Fallon.[35]

## JP-8 Jet Fuel

There is also the question of the jet fuel. Many residents of Fallon think that the 34 million gallons a year of JP-8 jet fuel that flows through town and is consumed by fighter jets at the air base may have been one of the contaminants that provoked the leukemia outbreak. There is logic to this reasoning. JP-8 jet fuel is made of benzene and kerosene, and benzene is not only a recognized carcinogen but has also been linked to leukemia. JP-8 also contains halogenated hydrocarbons and vinyl chloride, also potent carcinogens, notes Vera Byers, who studied a number of leukemia clusters.[36] Byers and her husband, Alan Levin, both of whom investigated the leukemia cluster in Woburn, Massachusetts, 1969–1981, believe that the jet fuel may have been one of the contaminants that caused the leukemia cluster in Fallon. Other studies, in which rats were exposed to JP-8 jet fuel, demonstrate that the substance is "linked to lung, kidney, liver, and DNA damage as well as suppression of the immune system."[37]

Witten is keeping an open mind on the subject. An expert on the toxic properties of JP-8 jet fuel, his research linked high doses of the substance to cancer in mice.[38] When mice were exposed to jet fuel in laboratory tests, their white blood cell counts did not become elevated. Since leukemia is a disease in which a damaged white blood cell begins to replicate uncontrollably, pumping out so many damaged cells that it overwhelms the body's ability to fight infection, the low level of white cells in the laboratory mice exposed to jet fuel seemed a counterintuitive result. Some commentators speculated that the experiment proved that jet fuel could not have caused the leukemia cluster. However, Joe Wiemels, the UCSF genetic researcher, points out that the best-known leukemogen, benzene, a chemical contained in JP-8 jet fuel, does not raise white blood cell counts and in fact depresses them. Benzene was actually used as a treatment for leukemia in the twentieth century before it was linked as a cause of the disease, he notes.

Although human exposure to JP-8 jet fuel has not been proven to cause leukemia, there are enough data about components of it, particularly benzene, to make it a plausible culprit. But proving that it might have been one of the chemicals that caused the cluster requires demonstrating a possible route by which the children who fell ill with leukemia might have been exposed. Here, several possibilities present themselves, including leakage of jet fuel from the Naval Air Station at Fallon, leakage from the Kinder Morgan Energy Partners jet fuel pipeline that runs through town, and dumping of jet fuel by fighter jets prior to landing.

### Contamination from Naval Air Station

Military air bases are notorious sites for contamination. The Naval Air Station at Fallon is no exception. The 7,982-acre main station at the base, located six miles southeast of Fallon, is dotted with areas contaminated with jet fuel and other chemicals. The base, through which some forty thousand military personnel pass every year, is littered with twenty-six toxic waste sites, sixteen of them contaminated with jet fuel.[39] State officials from the Nevada Department of Environmental Protection also identified twenty-seven contaminated sites on the base that warranted further investigation. Several plumes of jet fuel and other hydrocarbons are located beneath the surface of the base. In addition, naval fire control specialists deliberately set jet fuel and napalm on fire on the ground as part of their training exercises.

Regulatory officials have required the installation of numerous monitoring wells to see if pollution on the base is migrating off it in the groundwater. A report by ATSDR found little cause for alarm in terms of the Naval Air Station's impact on public health. In a public health assessment, the agency found that contaminated groundwater at the base does not threaten public drinking water, private wells, or surface water.[40]

Despite these reassurances, there have been jet fuel spills at the base over the years, and "about 65,000 gallons of contaminated fuel have been pumped from the ground since 1993," notes Reno reporter Frank Mullen. Jeffrey St. Clair, an investigative journalist, claims that more than this is missing: base officials are unable to account for more than 350,000 gallons of jet fuel despite the fact that federal officials required them to put in place an oversight system to keep track of the fuel in 1989. Navy officials claim that they lose only an average of 45 gallons a year in spills but whistle-blowers allege that they know of 30,000 gallons that leaked from the pipeline and a truck in 1988 and 1989. Also, two underground fuel storage tanks, each with a capacity of half a million gallons, are corroding due to contact with underground saltwater. Neither of the tanks is equipped with leak detection or overflow protection devices.

### Raining Jet Fuel

Some Fallon residents claim they are periodically exposed to jet fuel as it rains down on them when jets jettison fuel before landing. Officials at the Naval Air Station, however, claim that the only time their pilots dump fuel is during an emergency, something that happens rarely. The rest of the time what observers on the ground are seeing is water vapor created as fighter jets flying at supersonic speeds squeeze the water out of the air, notes Zip Upham, a communications official from the base.[41] Basing its conclusions of information provided to it by officials at the Naval Air

Station Fallon, the ATSDR concluded that the jettisoning of fuel on the 41,200 take-off and landing cycles at the base that occur annually do "not pose a past, current or future public health hazard."[42]

But this explanation does not pass muster with some residents in Fallon who report witnessing aerial dumping of jet fuel and later finding an oily residue on their cars and houses. Some residents claim the jets regularly dump fuel when they are traveling at low air speeds prior to landing. Cattle and horse ranchers are also familiar with the practice. One resident who runs cattle on a pasture near the base said that jet fuel dumped by fighter jets had coated the hides of her cattle, causing them to crack and peel off, killing the animals. "They were skinned alive," she reports. A horse rancher confirms the story: his horses experienced similar problems and died.[43]

## Kinder Morgan

Another possibility is that Fallon residents may be exposed to fumes from JP-8 jet fuel as a result of leakage from the Kinder Morgan Energy Partners forty-five-year-old pipeline. The pipeline follows the railroad tracks for part of the way through Fallon before jogging through town past a number of schools on its way top the Naval Air Station. Both residents and medical researchers have noted that a number of the families whose children fell ill with leukemia lived in housing less than a third of a mile from the pipeline. Numerous residents report strong petrochemical odors and have witnessed repairs of the pipeline, sometimes carried out at night. "We occasionally smell a kerosene-like odor. We get a whiff of it now and then. Teachers mention it to me," Scott Meihack, the principal at the E. C. Best School in Fallon, told a reporter from Reno.[44]

Officials at Kinder Morgan say that leaks from its pipeline are relatively minor and that they have not contributed to any environmental contamination that might have caused the leukemia cluster in Fallon. As evidence they can point to a clean bill of health report they received from the ATSDR, which, in a report released in 2002, stated that the jet fuel pipeline never leaked, is not leaking, and does not pose a threat to the health of residents.[45]

But the ATSDR never did any of its own sampling, notes Floyd Sands. Instead its findings are based on a study done by contractors paid by

Kinder Morgan Energy Partners who used tracer chemicals to detect leaks. Pipeline officials say that they use pressure tests to see if the pipeline is leaking and conduct aerial inspections of the line on a regular basis. In 1987 and 1997, the company also sent a robotic surveillance device through the line looking for cracks, but none were found.

These inspection techniques, however, are not foolproof. A hairline fracture in a pipeline later purchased by Kinder Morgan leaked sixty thousand gallons of fuel into the ground before it was discovered two months later when residents complained about the smell. Soil testing along the line where it passes through Fallon has not been conducted, nor has one of the wells closest to the line ever been tested, notes Sands.

Many residents of Fallon remain unconvinced by the ATSDR report. The Naval Air Station expanded in the mid-1990s, about the time that the spike in tungsten and jet fuel was found to have peaked in tree core samples, notes Brenda Gross, the parent of a child in the Fallon leukemia cluster. That year also marked the beginning of the leukemia cluster, she adds. Gross does not believe the assertion of pipeline officials that there were no spills. Residents in town witnessed pipes being dug up and patched, she says: "People have seen them repair the leaks. They say it's maintenance, but it is still a leak. We've seen them lift up and wrap the pipes before putting them back. I think they should test the soil along the pipeline every hundred feet. And they shouldn't have the pipeline company do the testing as they did. It should be another company to avoid a conflict of interest."

Kinder Morgan has demonstrated "widespread failure to adequately detect and address" both corrosion and outside force problems that cause leakage, writes by Stacey Gerard, associate administrator for pipeline safety at the U.S. Department of Transportation in a "corrective action" memo. "This failure has systematically affected the integrity of [Kinder Morgan's] Pacific Operations unit." As a result there were forty-four pipeline accidents from 2003 to 2005, including accidents in which petroleum products were spilled, among other places, in Fairfield, Carson, Martinez, San Bernadino, and Truckee, California; Fort Bliss Military Reservation in El Paso, Texas; and a gasoline spill in Walnut Creek, California, that killed five workers.

## Groundwater Contamination

Could the leukemia cluster have resulted from groundwater contamination? U.S. Geological Survey (USGS) tests of groundwater quality conducted in 1994 detected radioactive contaminants that exceed current standards in thirty-one of seventy-three wells tested. This study was not considered during the investigation into the cause of the Fallon leukemia cluster until 2001, when a former USGS director, John Nowlin, brought it to light. The study found that naturally occurring uranium coming from the Sierra Nevada mountain range had migrated into Fallon, affecting wells that drew from the shallow and intermediate aquifers. (Unaffected were the municipal wells, which supply a third of the town, because they draw from the deepest aquifer.) One of the wells sampled contained 310 micrograms per liter of uranium, and another had 210 micrograms per liter. In all, nine of the fifty-six wells sampled showed levels of uranium that exceed the EPA standard of 30 micrograms per liter promulgated in 2000. This was of interest to investigators because radiation is one of the few known triggers for leukemia.[46] However, testing of the water in the homes of families with children in the leukemia cluster found only one well with high levels of radiation.[47]

In 2007, another USGS study reported that twenty-five drinking water wells in Fallon were contaminated with polonium-210, a radioactive material known to cause cancer in humans. Thirteen of the twenty-five wells tested had concentrations of polonium-210 that exceeded the EPA's maximum content levels for gross alpha radioactivity in public supply wells.[48] (The EPA has no specific standard for polonium-210.) The worst-contaminated wells had 67.7 pico curies per liter of the radioactive isotope, some four times the EPA's 15 pico curies per liter standard.[49] Chris Pritsos at the University of Nevada at Reno is exposing laboratory rats to a combination of tungsten, arsenic and radioactive polonium-210—all contaminants found in the Fallon water supply—to see if these exposures cause genetic damage.[50]

## Arsenic and Mercury

Also detected at high levels in Fallon's groundwater and sediments was arsenic, a naturally occurring element that is found in the earth's crust at average background concentrations of 2 to 5 ppm. Levels of the metal in Fallon were "elevated compared to the national average," an ATSDR report noted. In fact, in seventy-two samples taken in Churchill County, including twenty-two in Fallon, the average concentration was 81.1 ppm and the highest concentration was 680 ppm. Although these levels far exceed the agency's environmental media evaluation guides (for a child, 20 ppm, and for an adult, 200 ppm), ATSDR nevertheless concluded that these levels were safe.[51]

Arsenic, found at 10 ppm in Fallon's drinking water, was subsequently reduced to 3 ppm once a water treatment plant was opened. Outside Fallon, where people drink their own well water, levels of arsenic have been found as high as 2,200 ppm. Some experts believe that exposure to these concentrations of arsenic could have triggered leukemia among Fallon's children

Mercury contamination is an issue as well. Fallon lies in a basin that experiences a drought-flood cycle. Gold and silver mining operations in the 1800s discharged an estimated seventy-five hundred tons of mercury into the Carson River drainage basin. From 1994 to 1996, there were droughts in Fallon, and then the flood came in 1997. Huge amounts of water washed down into the irrigation canals (where children sometimes swim), and excess water was permitted to flood out into the fields around town. The floodwaters carried sediments from Lake Lohattan, located eighteen miles west of Fallon, which is known to contain sediment with high levels of mercury.

The ATSDR conducted sediment sampling and found an average of 323 ppm of mercury, with the highest concentration at 13,100 ppm. Residents are most likely to come into contact with these contaminated sediments as a result of work or recreational activities and by eating fish or wildlife. Fish sampling found 0.02 ppm to 11 ppm wet weight in fish tissues, while the safety standard is 1 ppm. The average mercury concentration in duck muscle and liver was 5.9 to 17.8 ppm. The FDA issued a health advisory for fish and duck caught in the area. And the ATSDR warned that "human consumption of mercury-contaminated fish and duck poses a potential health hazard, especially for young children and pregnant women for long-term exposures."[52]

Could some of Fallon's children who developed leukemia have been exposed to some of these mercury-contaminated sediments in the flood control and drainage canals? asks Brenda Gross. To her knowledge, no one has investigated the irrigation ditches and fields to find out, despite the fact that children are known to have waded and played in the ditches. Faced with this possibility, the ATSDR recommends that "concerned persons could reduce the potential for exposure by cleansing skin and washing clothes after contact with surface water and sediment."[53]

## Nuclear Testing

Finally there is the question as to whether radioactive fallout from the testing of nuclear weapons may have played a role in Fallon's leukemia cluster. At the height of the Cold War in the 1960s, military experts tried to determine whether the Russians could mask a nuclear test detonation and claim that it was an earthquake. To test this theory, the Shoal Vela Uniform Project was initiated, and a 13 kiloton nuclear bomb was detonated underground in 1963 in the Sand Mountain range twenty-eight miles east of Fallon. Department of Energy officials assert that no radioactivity from the test site reached Fallon and that groundwater flows away from town and feeds a different aquifer.[54] But this was hardly the only nuclear test in the Nevada desert that could have produced fallout. Before the effects of radiation were fully appreciated, some Fallon residents report sitting in a circle next to the test site to witness an underground detonation. Other long-term Fallon residents claim they watched the mushroom clouds rise from the desert floor during above-ground tests. As a result, tritium can still be found in the water in huge craters that dot the desert floor. Whether radiation has anything to do with the more recent leukemia cluster in Fallon is unknown.

"I do feel that the leukemia cluster was caused by a combination of environmental contaminants," says Brenda Gross, whose son was in the Fallon cluster. Among the potential contributors she lists tungsten and cobalt from local manufacturing, jet fuel, arsenic in the water, mercury from past gold and silver mining activities, and fallout from radioactive testing.

While much attention focuses on environmental contaminants as possible sources of Fallon's leukemia problem, some experts believe it might have been caused by a virus and point to the population-mixing hypothesis, a theory based on a study of a childhood leukemia cluster near a nuclear power plant in Great Britain. It posits that elevated levels of leukemia found in remote rural areas are caused by an increase in visits from outsiders who bring with them viruses to which the local population is unaccustomed. These viruses then create susceptibility to disease among vulnerable members of the population. Chromosomal damage from the infection could lead to cell abnormalities that cause leukemia.[55] Mark Witten at the University of Arizona notes that leukemia clusters in both Fallon and Sierra Vista were preceded by outbreaks of shingles in children, a virus-borne disease usually found in adults.[56] However, Sands finds the population-mixing theory weak, noting that there was no significant change in the influx and outflux of outsiders prior to the disease cluster in Fallon.

### Precautionary Principle

Although no conclusive evidence has yet been found that proves that the leukemia cases in Fallon were caused by environmental contaminants, the evidence is sufficient to warrant that some remedial steps be taken. A precautionary approach to the contamination found in Fallon might require some combination of the following steps: rural residents dependent on wells with high arsenic levels be provided with safe drinking water; irrigation ditches be monitored for mercury content; further action be taken to limit resident exposure to tungsten-laden emissions; and further monitoring of radioactive pollutants in the water supply be carried out. In addition, an aggressive cleanup of Fallon Air Base should be completed, and the practice of aerial dumping of jet fuel should be limited to emergencies. Finally, an impartial third-party investigation into possible leaks in the jet fuel pipeline should be undertaken.

The Fallon case suggests that public health officials need to be more proactive in ensuring that residential populations are not exposed to heavy loads of multiple contaminants. Since neither medical nor environmental sciences are advanced enough to detect what causes disease clusters such as the one identified in Fallon, it makes sense to follow the precautionary principle in these hot spots of contamination. In the future, we will likely develop better tools and an expanded knowledge base to solve the puzzles posed by disease clusters. In the meantime, we need an improved approach to investigate and help residents in communities with disease clusters, we need more timely and targeted collection of both morbidity and mortality data so that we can spot problems early, and we need to minimize toxic exposures.

More specifically, the experience of residents in Fallon with the ATSDR's disease cluster investigation suggests that the agency's approach is in need of serious overhaul. New approaches must be developed that permit federal investigators to move quickly to identify potentially dangerous chemical exposures and work to alleviate them. This approach can be initiated prior to scientific proof of a public health threat being established. Enough is known about the toxic effects of tungsten, arsenic, mercury, jet fuel, and the radioactive pollutants found in Fallon that a case can be made that the synergistic effects of these pollutants could cause disease in ways that we do not yet fully understand. Demanding that state and regulatory agencies put into practice precautionary public health measures does not preclude the possibility of continuing research that follows strict scientific methodology.

# Conclusion

Sacrifice zones are a blight on the land. They bear witness to an ongoing and pernicious form of racial and class discrimination that this nation has yet to address.

The dozen sacrifice zones you have just read about provide evidence of a pattern of environmental injustices in which low-income and minority populations are at greater risk of being exposed to health-destroying toxic chemicals than are wealthier and better-protected Americans. This collection of personal testimony from hundreds of residents in fenceline communities is backed up by numerous academic studies that have quantified the disproportionate toxic burden that residents in these hot spots of pollution experience.[1]

For example, a review of the U.S. Environmental Protection Agency's Toxic Release Inventory (TRI) data revealed that people living near heavily polluting industries are disproportionately exposed to toxins. This should come as no surprise given that, nationwide, 93.9 percent of the twenty-three thousand largest polluting facilities release their pollutants on site into the air, water, and soil.[2] "Thus citizens who work and reside in the areas in which these facilities are located typically experience much greater rates of exposure to industrial pollutants."[3]

Despite the clear findings of this accumulating body of research, how close a person lives to a heavily polluting industry remains a largely overlooked measure of inequality in the United States. Commenting on the basic unfairness of this division of environmental burdens, Benjamin F. Chavis Jr. noted that the environmental justice movement "confronts the immorality of upper- and middle-class people consuming the most energy and producing the most waste, while it is the health of the poor that is most affected by the resulting pollution."[4] This sentiment is shared by

Robert D. Bullard, one of the most knowledgeable writers about environmental justice issues who observes that "low-income and minority communities continue to bear greater health and environmental burdens, while more affluent whites receive the bulk of the benefits."[5]

## Why Do We Permit Unequal Exposure?

Why are some citizens exposed to intense pollution while the vast majority of Americans can afford to avoid this contamination? Part of the problem can be attributed to the fact that our regulatory system is not geared to protect residents who live near heavy industries. Environmental regulations do a halfway decent job of protecting the majority of Americans who live at a distance from intense sources of pollution; but they do little to safeguard the well-being of those who live in the shadow of heavy industry and cannot afford to move to a safer location.

Some government officials have acknowledged this glaring deficiency in our environmental protections that fail to adequately regulate industrial pollution near residential areas. In introducing the environmental justice bill in 1992, which failed to be enacted into law, former Vice President Albert Gore Jr. observed that the United States "faces disturbing inequities in the way severe pollution problems are distributed. . . . In disproportionate amounts, toxic wastes and toxic emissions from industrial processes contaminate the neighborhoods of minority communities."[6]

While environmental justice scholars have convincingly demonstrated that high-emission plants and noxious public utilities are disproportionately sited in low-income communities of color, where they create illness and lead to early death, their persuasive argument calling for long-overdue regulatory relief for these neighborhoods has to date been largely ignored by the mainstream American political apparatus.

True, while he was president, Bill Clinton issued executive order 12898 on February 11, 1994, which required all federal agencies to consider environmental justice factors when making decisions. Despite this initiative, new toxic industries are built every day in already overburdened and overpolluted sacrifice zones across the country. Reports produced by the Government Accountability Office, the EPA's inspector general, and the U.S. Commission on Civil Rights all conclude that the EPA has failed to integrate environmental justice considerations into its decision-making process or made them a core part of their mission.[7]

So why do we not pass legislation to better protect fenceline residents. Why are we as a nation all but deaf to the environmental justice argument? Why are the general public and the press not calling for reform? Why do we allow a form of discrimination to persist in which toxic exposure burdens are clearly apportioned by race and class? And why do we sit idly by as the most toxic industries are sited in low-income and heavily minority communities?

There are no comfortable answers to these questions because sacrifice zones do not exist by accident. They are in fact shaped by powerful forces. Some will argue that legislation designed to reduce environmental inequities will never pass Congress because the corporate lobby is too strong. That the environmental justice legislation failed to pass Congress is evidence of this. To date, the pro-industry arguments outlined below have dissuaded members of Congress from enacting laws that might provide relief for fenceline residents.

Industry lobbyists argue to members of Congress that installing state-of-the-art pollution control equipment in industries near residential areas would be costly and would raise the price of many consumer products. This argument, however, fails to take into account studies demonstrating that improvements in environmental conditions are associated with both rising wages and property values. Despite these findings, large industries have been remarkably effective at lobbying Congress for the passage of laws that permit them to externalize their costs by forgoing investments in pollution control technologies. This license to pollute keeps corporate costs low and American jobs from being exported, industrial lobbyists contend.

The problem with allowing heavy industry to emit large volumes of toxic chemicals into residential neighborhoods is twofold. First, permitting industries to externalize their costs simply transfers them to residents of sacrifice zones who breathe in contaminated air and drink polluted water. Their health is impaired from exposure to these elevated levels of chemical pollutants, and they pay the cost in terms of suffering and early death. But taxpayers end up paying for the treatment of many of these uninsured fenceline residents who flood into our emergency care system.

A further problem with the argument that insists that American industry can be competitive in a global marketplace only if it is permitted to forgo investments in the best available pollution control technologies is that it retards the development of inventions that will protect both the environment and public health. These innovations will lead to products and processes that use less energy and fewer toxic chemicals. Stricter regulations, better enforcement, and higher fines would provide incentives for industry to invent less toxic alternatives to current practices. And these new alternative technologies are precisely the ones that are likely to be in ever greater demand as regulatory regimes around the world become stricter over the coming decades.

The Dickensian conditions in sacrifice zones, described in detail in the previous chapters, exist for political as well as economic reasons. In order to win election to public office, candidates need money to fuel their campaigns, and most of them accept contributions from big industries. When votes come up in Congress and in the state legislatures about what kind of pollution control regulations to pass and how laws are to be enforced, lobbyists for these top industrial donors call in their chits by demanding relaxed emission standards. The results of these lobbying efforts frequently shortchange the interests of low-income, relatively powerless residents who live next door to industry.

Bring up the subject of environmental injustice with people who are skeptical about the government's ability to solve problems such as this, and you are likely to hear myriad excuses about why we cannot afford to do anything about it. Some argue, for example, that tightening environmental regulations to reduce emissions will only drive more companies to export manufacturing jobs to developing nations, where rules are less stringent. This argument has the merit of being logical, but it fails to explain why we feel justified in allowing low-income people of color to be exposed to disproportionately high concentrations of toxic chemicals while the health of more affluent whites is protected. If sacrifices must be made for the greater economic good in a democratic republic, then surely they should be evenly shared.

A less commonly voiced (but nevertheless widely held) perspective is that environmental injustices are morally repugnant but too expensive to remedy. Those who hold this view contend that in most cases, relocating populations away from the fenceline with heavy industry is not a viable option because the large number of residents in these areas makes this approach prohibitively expensive. They further argue that the United States should not be a "nanny state" and individuals who currently inhabit sacrifice zones should protect themselves by moving away from sources of heavy pollution. "Let the market work this out," reason those who believe that people who do not like the pollution load where they live should just pack up and move elsewhere. This perspective fails to take into account that most residents of sacrifice zones either do not have the money to relocate or refuse to abandon family and friends who cannot afford to leave.

Another reason that environmental injustices have been largely ignored by most Americans is that the issue does not hit them where they live. Our residential landscape has become so stratified by income group that middle-class and affluent Americans rarely have occasion to visit fenceline communities. As a result, they have no clue what life is like in them. Furthermore, those who downplay the public health problem posed by sacrifice zones tend to naively believe that regulations and enforcement measures are in place across the nation that protect all citizens from toxic industrial emissions; and that state and federal regulators are dealing effectively with this problem. They argue that their government does at least a halfway decent job of protecting its citizens from industrial contaminants and certainly a better job than is done in many other nations. And they point out that since the Clean Air Act was passed in 1970, the quality of the air has improved overall. Similarly, there have been improvements in water pollution controls, and rivers no longer catch fire as they did in earlier decades. Despite this overall progress, a segment of the American population has been left behind in pockets of heavy pollution.

## Civil Rights Issue

Cleaning up environmental conditions in fenceline communities across the nation would vastly improve the lives and protect the health of millions of Americans. Accomplishing this ambitious goal, however, will require a sustained effort from a broad coalition of forces. What follows are some ideas about how to make progress on this issue.

Above all, the problem posed by the existence of sacrifice zones is a civil rights issue and is best understood and addressed as such. "Equal justice and equal protection from pollution" is the goal of environmental justice movement," points out Robert Bullard, one of the movement's founders. In his essay, "Social Construction of Environmental Justice," Stephen Sandweiss takes this analysis one step further: "By incorporating environmental concerns into a civil rights frame, the environmental justice movement was able to characterize the distribution of environmental hazards as part of a broader pattern of social injustice, one that contradicts the fundamental beliefs of fairness and equity."[8] Describing the environmental justice movement as a logical extension of the civil rights movement also allows environmental justice activists to "draw upon the organizational resources and institutional networks established during the previous struggle for racial equity," he continues. Historically black colleges and universities, black churches, and neighborhood improvement associations "furnished the environmental justice movement with leadership, money, knowledge, communication networks and other resources essential to the growth of any social justice movement."[9]

## Integrated Approach Required

Environmental injustices in sacrifice zones around the country must also be viewed as a public health problem that traditional approaches have failed to remedy, as well as a misguided land use planning policy that has resulted in the incompatible juxtaposition of residential and industrial zones

The problems facing residents in fenceline communities are multifaceted and "comprise a complex web of public health, environmental, economic, and social concerns. Given the multiple stressors that impact low-income, people of color, and tribal communities, such groups do not have the luxury of addressing one issue at a time," observes Charles Lee, chair of the EPA's Office of Environmental Justice.[10] Not only are residents in sacrifice zones exposed to disproportionately high levels of toxins, Lee continues, many of them are also unusually susceptible to ill effects from these exposures due to compromised immune systems, poor diet, lack of access to medical care, and a host of other problems from which low-income people suffer. Improving the health of this population, he writes, will require a "holistic, community-based, multi-issue, cross-cutting, independent, integrative approach," which has as its goal the creation of "healthy, livable, sustainable, and vital communities."[11]

A consensus is gradually emerging among scholars that solving environmental justice problems will require a community-based, participatory approach, capacity building within affected communities, and partnerships that permit fenceline residents to use and leverage the resources of the government. These efforts will also demand a more sophisticated level of preventive public health measures than those deployed today and improved land use planning that is more protective of public health. In other words, solving the problem posed by environmental injustice will entail not just that government officials understand the problem and agree that intensive efforts should be made to solve it, they must also push interagency cooperation to a level rarely achieved. To accomplish this, national leaders will need to be educated about environmental injustices through a national dialogue around the issue.[12]

This ambitious agenda is likely to take years, if not decades, to fully implement. Deploying the full resources of our public health agencies in fenceline communities would be a major accomplishment. Tightening enforcement of emission regulations will take time. And convincing local zoning boards to avoid financially attractive land use decisions that place fenceline residents at risk of harm from industrial pollutants will be, if it is ever achieved, a breakthrough. Nevertheless, this is the scope of the work that needs to be done.

Some of this work has already begun. The Interagency Working Group on Environmental Justice, established by President Clinton's executive order 12898, arrived at the Integrated Environmental Justice Action Agenda, which spawned fifteen environmental justice demonstration projects. These projects are designed to show "how land-use planning, pollution prevention, use of new technologies, and capacity building are being applied to the context of the holistic vision of community revitalization," notes Lee.[13]

On the ground this has meant "turning environmental liabilities into community assets and opportunities" by linking community cleanup, transformation, and revitalization into a source for local jobs, observes Mary Nelson at Bethel New Life, a faith-based community development group in the largely African American neighborhood of West Garfield in Chicago. Similarly, in the South Bronx in New York City, community-based organizations identified the stationing of vehicle fleets in their neighborhood as a source of the air pollution that was causing elevated rates of respiratory disease. To help alleviate the problem, the U.S. Postal Service agreed to spend $1.93 million purchasing fifty-five electric and natural gas vehicles.[14] These and other projects are beginning to create partnerships between grassroots fenceline groups and government officials to address some of the root causes of environmental inequities.

### Nuts-and-Bolts Remedies

What other remedies are available to eliminate this systemic violation of civil rights in fenceline communities? The first and most obvious step is to stop siting heavily polluting facilities in already overburdened fenceline neighborhoods. As one expert writes, this will require "the establishment of local state and federal government programs and policies that ensure environmental equity; avoid the siting of hazardous facilities/sites in already overburdened low-income communities and communities of color; provide resources to these overburdened communities to create environmental amenities that can partly offset other environmental risks; and promote greater citizen participation in the problem-solving and decision-making process that affect those communities."[15]

Additional environmental protections and targeted enforcement actions are also needed to safeguard the health of fenceline residents. Existing laws have proven inadequate to deal with these hot spots of industrial contamination. Furthermore, federal and state environmental regulatory agencies have done a poor job of enforcing existing regulations to protect public health in fenceline communities. The record of state regulatory lethargy and unwillingness to take strong emission enforcement actions along the fencelines in Texas, Florida, Louisiana, and Ohio is particularly troubling.

Much could be done to reduce toxic exposure in neighborhoods near industry by enforcing existing pollution emission laws. Some experts are now calling for the creation of special regulations for what they term areas of critical environmental concern. These fenceline areas will require that "higher scrutiny in environmental permitting and greater levels of resources for cleanup and remediation" be deployed in order to create improved environmental equity, argue Daniel R. Faber and Eric J. Krieg.[16]

Better enforcement of air and water pollution laws near major industries, however, will be meaningful only if it is backed up by substantial fines and a willingness to close down companies that repeatedly violate pollution emission laws. Otherwise large industries will continue to regard the slap-on-the wrist fines that are currently assessed as an affordable cost of doing business. Furthermore, monies garnered from these pollution fines should be reinvested in the communities where the chemical trespass has taken place. These monies could make it possible for local watchdog groups to buy sophisticated monitoring equipment and pay for health surveys to better make the case for regulatory relief.

Regulatory agencies must also be prepared to take action in cases where industrial facilities show a pattern of emitting large volumes of toxic chemicals beyond their permit. Many such releases by industry are currently excused as unpredictable accidents or upsets. Regulatory agencies should scrutinize these apparent accidents for a pattern that makes them both predictable and avoidable. If some of these upsets do prove predictable, fines should be assessed until the systemic problem is remedied.

Ultimately companies that persistently violate their air or water pollution permits after repeated warnings and fines should be given a choice of installing pollution control equipment that effectively abates the problem, creating an adequate buffer zone, relocating their plant, or paying to relocate the residents they have placed at risk in homes that are comparable to or better than the ones they currently inhabit. If facility owners and managers balk at all of these options, their facilities should be shut down.

## Reducing Toxic Feedstocks

In addition to the stick approach to stricter enforcement, there should also be a bunch of carrots, or incentives, for industries to reduce their toxic emissions. Our tax system could be used to deliver these incentives by substituting some taxes currently levied on individual income in favor of a toxics tax targeted on goods made with toxic chemicals. This would render goods manufactured with fewer toxic inputs cheaper, giving consumers a price-feedback mechanism by which they could distinguish which goods are made in a way that protects public health and prevents environmental degradation. Such a tax would not only reduce the volume of dangerous chemicals emitted by industry; it could also play an important role in reducing the release of carbon dioxide and other heat-trapping gases that are causing climate change.

Although we should not use taxpayer money to pay industries to follow the law, industries should have an incentive to cut their costs by transforming their plants into cleaner-running, more energy-efficient facilities. If they do, not only would they spend less on energy and the purchase of health-damaging chemicals; they would also stand to avoid the need for additional costly pollution control technologies and fines. Finally, if businesses invent less toxic means of producing goods and services, environmentally conscious consumers may reward them by deciding to buy their products instead of less eco-friendly alternatives.

## New Monitoring Legislation

New legislation is also needed to protect residents of fenceline communities. Typically environmental regulations are not geared to deal with hot-spot problems where emissions are elevated by adjacent refineries, chemical plants, cement kilns, asphalt plants, tire manufacturers, steel and plastic plants, other industries that manufacture products from highly toxic substances, and military bases. New legislation should require high-emission industries to pay for specialized monitoring equipment capable of capturing and documenting emission violations along their perimeter. While the companies should pay for this equipment, they should not be permitted to collect, analyze, or send these emissions data to the appropriate regulatory agencies. These functions should be carried out by a neutral third party.

Since they live next door to high-emission industries, residents in fenceline communities are best positioned (and highly motivated) to provide frontline monitoring services and capture evidence of illegal emissions. All they need are the funds to purchase and operate mobile monitoring equipment. To date, few programs have provided resources to equip these residents to protect their health. Regulatory agencies should make available both funding and technical expertise for grassroots fenceline community groups so they can provide an early warning system that rapidly identifies dangerous pollution incursions into their neighborhood.

Finally, grassroots antitoxics activists need better laws to hold neighboring industries accountable for chemical trespass events, as well as access to lawyers trained in taking these cases to trial. Currently the law favors industrial polluters over the interests of fenceline residents because of the difficulty of proving that a specific release caused an illness. Laws need to be passed that do not require residents to prove that their health has been damaged by elevated releases of toxic chemicals but rather place the burden of proof on the emitter of these pollutants to prove that emission levels that exceed their permits have not caused illness.

## National Standards for Chemical Exposure

Most Americans assume that government regulators have uniform and nationally accepted standards for safe exposure levels for toxic chemicals. This is not the case. In many states there is a woeful absence of residential chemical exposure standards, and different agencies have varying regulations. This permits industry officials to argue that even their releases of large quantities of known carcinogens and other health-damaging chemicals do not endanger public health.

To compound the problem, all populations are not created equal when it comes to their ability to withstand environmental insults without serious health consequences. For example, chemical exposures can have a devastating impact on fetuses or infants during developmental windows of vulnerability when cells are dividing rapidly. Since infants and children have lower body weights than adults, they can be adversely affected by lower doses of toxic chemicals that adults can withstand. In addition, infants and children are particularly vulnerable because they are shorter than adults and thus closer to the ground where contaminants collect; they ingest more contaminants because they breathe faster than adults do and crawl in the grass and the dirt and put their hands in their mouths. Many chemical exposures regulations fail to take into account the special vulnerabilities of children and are geared instead to healthy adults. Other subsets of the population—such as pregnant women, the elderly, the chemically sensitive, and the sick—may also be disproportionately affected by exposure to certain pollutants and should be taken into account when promulgating standards for safety.

Since we have very few data about the health impacts of many of the new chemicals being introduced into industrial processes, consumer products, and agriculture, regulators should follow the precautionary principle. One would think that chemicals that had not been tested for their potential health impacts would not be used in large volumes prior to testing, but this is not always true. As a result, fenceline residents are frequently unwitting and unwilling participants in uncontrolled experiments in which they are exposed to untested chemicals that have the potential to compromise health. Our ignorance about the synergistic effect of multiple chemical exposures is even more profound, amplifying the need to be conservative about chemical exposure.

To help protect both worker and resident health, the U.S. Congress should establish nationwide chemical exposure safety standards that protect our most vulnerable populations so that everyone across the nation is protected. In addition to standardizing these regulations about safe exposure levels for individual chemicals, we must begin to put in place rules that limit the cumulative burden of exposure to multiple toxins.

**Health Surveys**

One of the most dismaying aspects of sacrifice zones is that lack of interest they receive from public health officials. Many of these fenceline communities are experiencing a public health emergency, yet they are rarely visited by public health officials. Why the public health field has ignored the problems of this vulnerable population, under daily assault by toxic environmental emissions, is a worthy subject for investigation. Public health officials need to do more research and provide medical help for fenceline residents. Health surveys should be carried out regularly in sacrifice zones where contamination problems are most likely to cause disease. Currently there is an appalling absence of public health data about disease rates in fenceline communities. Annual health surveys and health monitoring in high-risk communities on the border with heavy industry could help catch environmentally induced outbreaks of disease before they become epidemic. The new technique of body-burden monitoring—in which blood, urine, hair, and tissue samples are analyzed for toxic chemicals and heavy metals—could prove a valuable tool in determining changes in exposure levels among residents.

Health studies in sacrifice zones should be designed to capture the full extent of disease in the community. To this end, residents should be questioned about their health and the health of members of their family (or former residents) who have already died or moved away. In this fashion, a time line of the historic health problems in the community can be assembled that can aid in targeting health protection programs.

## Media Coverage

More media coverage and public education programs about the peril that residents in sacrifice zones face are needed to pressure regulatory officials into protecting the health of residents in all parts of the country, including in low-income and heavily minority fenceline neighborhoods. Residents who stage protests and attract the attention of the media can improve regulatory enforcement. Most corporations care deeply concerned about their image (and brand name) and understand that in a competitive market, their profits may be reduced substantially if they earn a reputation for not caring about how their emissions cause disease among neighboring residents. This sensitivity about a corporation's environmental record gives local citizens leverage even with the largest corporations. Why the mainstream media have been slow to fully air the nationwide problems facing residents of fenceline communities is another question that begs for attention. Is it because publishers and editors are concerned about overwhelming readers with bad news, do they worry about appearing to be antibusiness, do they fear lawsuits, do they want to avoid alienating potential advertisers, or is it some combination of all of these factors?

## Buffer Zones

Another step toward alleviating the problems experienced in sacrifice zones would be to establish buffer zones—"breathing space"—that would mandate a safe distance be maintained between heavy industry and residential developments. Ideally heavy industry should be at a safe distance from residential populations. A greenway around high-emission plants can help distance residents from the worst exposures to toxic emissions. Questions remain, however, as to how large these buffers need to be, and what should be done in existing locations where large numbers of residents are already living near heavy industry. Clearly, choosing which sites most desperately need buffer zones will require prioritizing based on risk. But while we deal with the high-priority cases, we should also be adjusting long-term land use planning and zoning rules to avoid the creation of new sacrifice zones.

Ideally, heavy industries and waste sites should be surrounded by buffer zones and then light industrial zones occupied only by employees who work in protected structures or wear safety equipment. The next concentric circle out might include commercial operations where people work only eight hours a day, thus limiting their exposure. Beyond this could be the residential areas, followed by a more distant zone for schools, hospitals, and day care centers. The European Union is already experimenting with this concept; we should follow its lead given that it will take decades to rearrange the built landscape along these lines.

Skeptics about the possibility of profound political change no doubt will warn that the adoption of a land use plan this rational is unlikely to happen anytime soon. Local jurisdictions jealously guard their zoning prerogatives, making the nationwide mandating of buffer zones difficult to achieve. But the political climate is always susceptible to organized pressure, and it is not unreasonable to expect that, in the near future, a coalition could be assembled to reduce the number of new sacrifice zones being built every day by requiring high-emission industries to purchase properties that are large enough to include substantial buffer zones. How large should the buffer zone be? It should be large enough so that any pollution from the plant is attenuated to safe levels by the time it reaches residential areas.

## Brownfields and Sacrifice Zones

The problem that disproportionately contaminated communities face is not new, nor are government efforts to clean them up. Over the past several decades, the federal government has created programs to clean up thousands of contaminated sites around the nation left behind by industries that polluted the land before going bankrupt or moving away. There were so many of these brownfield sites around the country that federal officials recognized the need for a systematic approach to decontaminate them.

One of the persuasive political arguments for reclaiming brownfield sites (in addition to the health threat they posed) was that they needed to be cleaned up because they hampered economic growth: no one wanted to buy a house or business next to a contaminated property. As a result, property values in brownfields declined, and large areas of the country became disinvestment zones with boarded-up businesses and homes. The appropriation of funds to clean up brownfields was ultimately sold to Congress as an effective economic stimulus initiative. Clean up the contaminated property, and someone will buy the land; the property will go back on the tax rolls and generate revenue that can be used to pay for municipal services and infrastructure improvements. While much remains to be done to continue the work of brownfields cleanup, the program has raised a lot of money to do this work, brought thousands of properties back onto the market, increased tax revenues, put people to work rebuilding their communities, and continued for years with bipartisan support. This raises the question: Could such a strategy work for sacrifice zones?

One of the outstanding differences between brownfields and sacrifice zones is that while abandoned properties can be decontaminated, the pollution coming from nearby industries is ongoing in sacrifice zones. As a result, attracting investment to these areas would require proving to prospective buyers that emissions from nearby plants and military bases could be controlled and that living in the neighborhood would be safe. While this is no easy task, if it is done effectively and large industries and military bases become responsible neighbors, then adjacent residential areas might undergo the same kind of economic resurgence that restored brownfields neighborhoods have experienced.

## Political Reforms

Beyond the practical considerations about how to improve environmental regulations, emission reductions, public health studies, and the creation of buffer zones are deeper issues raised by the existence of sacrifice zones that will not be solved by tinkering with the regulatory process. Some environmental justice researchers have pointed out "the complex dynamics of environmental inequality" and are beginning to focus on "disparities in political power" and the role of "residential segregation" in creating the problems that fenceline residents experience.[17]

In other words, to get to the root causes of environmental inequity requires confronting the fact that most of the residents who live in sacrifice zones are politically marginalized citizens who have little input into the decisions about the siting of hazardous facilities in their community and other issues that drastically affect their quality of life and health. Furthermore, fenceline residents live near heavy industry because that is where they can afford to live and because that is where low-income people of color were permitted to settle in past generations. As some environmental justice experts have observed, "Community participation is a key to developing long-term regulatory, enforcement, and regional development initiatives that are politically and economically sustainable and that protect public health."[18] Only political empowerment, they argue, will ultimately solve the problems posed by environmental injustice.

### Grassroots Groups on the Move

While many difficult challenges continue to face environmental justice advocates who want to clean up fenceline communities, some small-scale positive developments can already be reported. Around the nation, groups of residents in sacrifice zones are awakening to the threat to their health posed by heavy pollution from neighboring industries, getting organized, and beginning to protest the dumping of toxics in their backyards. These are first steps in a long campaign to reduce toxic exposures.

Furthermore, there have been some outright successes to celebrate in which resident activists in fenceline communities have forced companies that caused environmental problem to either shut down or relocate residents who live too close to the high-emission facilities. Elsewhere a number of grassroots groups have succeeded in requiring industry to invest in the installation of additional or improved pollution control technologies to reduce their emissions, pollution monitoring equipment, and improved emergency warning systems. Others kept heavy industry from moving into their neighborhood.

Among the techniques that have proved effective in improving conditions in sacrifice zones are political organizing; challenging industrial air and water pollution permits; waging media campaigns that embarrass industries into reforms; filing lawsuits; training and equipping sacrifice zone residents to do their own monitoring of air, water, and toxins in their own bodies; as well as launching public awareness campaigns, citizen audits of companies, and good neighbor campaigns.

As effective as these tactics have proven, small, underfunded, isolated, grassroots campaigns to clean up the air and water near heavy industries often do not have enough money or political clout to face the scale of this systemic problem. Ultimately larger forces will be required to prod government into protecting these populations. Eliminating sacrifice zones will require many years of effort by a new coalition of progressive forces: environmental justice activists, civil rights workers, union organizers, social justice advocates, mainstream environmentalists, and perhaps industrialists who want to operate in a responsible manner. Each of these groups will have to step up and place a higher priority on improving conditions in sacrifice zones.

### Mainstream Environmental Organizations

Convincing the leadership of the largely white, well-funded, environmental groups that they should make it a priority to support grassroots, fenceline, environmental justice struggles is critical. To date, the most powerful environmental groups (the big ten) have been largely missing in action, with some notable exceptions, when it comes to protecting the plight of the poor in sacrifice zones.[19] They have eschewed the small-scale, often contentious fenceline struggles in favor of megabattles over climate change and the protection of endangered species, forests, farmlands, open space, wetlands, and marine ecosystems. While all of these campaigns are necessary to create an ecologically sustainable future, mainstream environmental groups need to take seriously the needs of those in sacrifice zones; otherwise they risk dividing the larger environmental movement along race and class lines.

The environmental movement would become more racially and economically integrated and energized were it to support grassroots, fenceline, environmental justice struggles. After all, many residents of sacrifice zones are highly motivated to improve environmental regulation and enforcement. No one will advocate more forcefully for changes in air pollution and water pollution regulations that those who are drinking and choking on toxic industrial or military emissions. The top ten green organizations could help provide community groups in sacrifice zones with legal, financial, and technical support that would help them defend themselves against depredations by their giant corporate neighbors. A collective effort by these forces could marry economic and racial justice issues with the drive for an ecologically sustainable culture. This coalition could put a new face on the environmental movement in the United States and make it stand not just for conservation of nature but also for social justice.

Individual citizens and environmentally responsible businesses also need to do their part to ensure that environmental burdens are fairly shared in this nation. Each of us can send letters to newspapers and Congress about local fenceline struggles; vote for candidates who pledge to reduce environmental injustice; vote with our pocketbooks by supporting candidates and nonprofits that provide help to fenceline struggles; purchase nontoxic or reduced-toxic products; and refuse to buy stock in companies that have a dismal record on fenceline issues or buy a token amount of stock in the company and then advocate for change from within.

Finally, we must all be willing to look beyond the rationalizations and excuses for inaction and squarely face the fact that environmental injustices affect millions of citizens across the United States. We need to publicly acknowledge that the disparities in exposure to toxic chemicals follow race and class lines despite our cherished ethic that promises equal protection under the law. These uncomfortable facts reveal that not only do we Americans inhabit two separate and parallel worlds of the rich and the poor; we also live in a nation divided by pollution exposure levels. Only when these inequities in chemical exposure are acknowledged by millions of Americans will we begin to build the political will necessary to abolish these shameful environmental injustices so that all Americans are equally protected from industrial contaminants.

# Notes

## Introduction

1. In 1980 Congress passed the Comprehensive Environmental Response, Compensation and Liability Act, which provided the Environmental Protection Agency with the authority to identify parties responsible for hazardous waste sites that threaten public health. The EPA has the power to compel these parties to clean up the site; or, if no legally responsible party can be found, the agency can use specially designated federal funds to clean up the site. The EPA maintains a National Priority List of some forty thousand contaminated sites of which about a thousand are designated as priority sites for cleanup. Over the last decade an average of twenty-one sites per year have been designated as successfully cleaned up. Funds for a more rapid and comprehensive cleanup are lacking.

2. Keith Schneider, "Dying Nuclear Plants Give Birth to New Problems," *New York Times*, October 31, 1988.

3. Robert Bullard, *Unequal Protection: Environmental Justice and Communities of Color* (San Francisco: Sierra Club, 1994), 4.

4. Robert D. Bullard, "Overcoming Racism in Environmental Decision-Making," *Environment*, May 1994, 6.

5. Juliana Maantay, "Mapping Environmental Injustices," *Environmental Health Perspectives* 110, Suppl. no. 2 (2002): 162.

6. Paul Mohai and Bunyon Bryant, "Environmental Injustices: Weighing Race and Class Factors in the Distribution of Environmental Hazards," *University of Colorado Law Review*, no. 63 (1992): 927, cited in *Environmental Injustices, Political Struggles*, ed. David E. Camacho (Durham, NC: Duke University Press, 1998), 53.

7. David Pace, "More Blacks Live with Pollution," Associated Press, December 13, 2005, 1. http://hosted.ap.org/specials/interactives/archive/pollution/part1.html.

8. Daniel R. Faber and Eric J. Krieg, "Unequal Exposure to Ecological Hazards: Environmental Injustices in the Commonwealth of Massachusetts," *Environmental Health Perspectives* 110, Suppl. 2 (2002): 278.

9. Marianne Lavalle and Marcia Coyle, "Unequal Protection: The Racial Divide in Environmental Law," *National Law Journal*, September 21, 1992, S2–S12.

10. Stephen Sandweiss, "The Social Construction of Environmental Justice," in Camacho, *Environmental Injustices, Political Struggles*, 31.

11. J. Roque, "Review of EPA Report: Environmental Equity: Reducing Risk for All Communities," *Environment* 35, no.5 (1993): 29 , cited in Faber and Krieg, "Unequal Exposure to Ecological Hazards."

## Chapter 1

1. Christopher Curry, "Pollution Risks? Agency Finds What Could Be Violations at Royal Oak's Charcoal Plant in Ocala," *Ocala Star Banner*, November 30, 2005, 2, 3.

2. Ibid.

3. Christopher Curry, "Coat of Charcoal: Residents Concerned with Soot from Royal Oak," *Ocala Star Banner*, August 24, 2005, 1.

4. Curry, "Pollution Risks," 1.

5. Lashonda Stinson, "West Ocala Health to Get a Close Look," *Ocala Star Banner*, August 1, 2004, 1.

6. Christopher Curry, "Unkempt Properties, Pollution Cited in West Ocala Survey, *Ocala Star Banner*, April 10, 2005, 1.

7. Cara Buckley, "Air of Suspicion—Part 2—Unlikely Rebels Defeat Gritty Plant," *Miami Herald*, May 19, 2006, 1.

8. Cara Buckley, "Air of Suspicion—Part 1," *Miami Herald*, May 19, 2006, 2.

9. "Royal Oak: Company Remains Silent," *Miami Herald*, May 19, 2006.

10. Curry, "Unkempt Properties," 1.

11. Buckley, "Air of Suspicion—Part 1," 2.

12. Ibid., 3.

13. JoAnn Guidry, "Child by Child," *Ocala Style* (August 2007): 74.

14. Christopher Curry, "Soot Sleuths: Volunteers Monitor Industrial Plant Emissions," *Ocala Star Banner*, December 31, 2005, 1.

15. Buckley, "Air of Suspicion—Part 1," 1.

16. Curry, "Coat of Charcoal," 1.

17. Ibid.

18. Environmental Defense, "Scorecard: The Pollution Information Site," November 1, 2009. www.scorecard.org/chemical-profiles/.

19. Curry, "Pollution Risks," 2.

20. Christopher Curry, "Royal Oak Fined for Violations," *Ocala Star Banner*, August 1, 2006, 1.

21. Christopher Curry, "Royal Oak to Close Ocala plant: Plant to Shut Down in Face of Inquiry, Resident Complaints," *Ocala Star Banner*, December 2, 2005, 1.

22. Global Environmental Monitor, "Florida Community Triumphs over Royal Oak Charcoal Factory," 1, http://www.gcmonitor.org/article.php?id=512 &printsafe=1.

23. Curry, "Royal Oak to Close Ocala Plant," 1.

24. Ibid.

25. Christopher Curry, "Pollution Concerns Royal Oak Neighbors," *Ocala Star Banner*, February 1, 2006, 1.

26. Wilma Subra, "Soil Testing Results [Ocala]," June 6, 2007, 3, 4.Subra wrote this report at the request of the Neighborhood Citizens of North West Ocala (NCNWO) Ocala, Florida. It concerned the interpretation of the testing results of soil samples for Polynuclear Aromatic Hydrocarbons.

27. Curry, "Pollution Concerns," 1.

28. Subra, "Soil Testing Results [Ocala]," 9.

29. The U.S. Environmental Protection Agency administers a "brownfields" restoration program whereby some of the 450,000 contaminated sites around the nation may be eligible for funds to clean up contamination and thus make the sites attractive for redevelopment.

30. David E. Camacho, ed., *Environmental Injustices, Political Struggles: Race, Class, and the Environment* (Durham, NC: Duke University Press), 211.

## Chapter 2

1. Margaret L. Williams, "Environmental Justice: A Sacrifice Zone in Pensacola," Citizens Against Toxic Exposure, n.d., 2. For access to this report contact CATE at http://www.cate.ws/.

2. U.S. Environmental Protection Agency, "Superfund Proposed Plan Fact Sheet: Escambia Treating Company Site," August 2005, 1. http://www.epa.gov/Region4/waste/npl/nplfln/escwodfl.htm.

3. Beyond Pesticides, "Poison Poles: A Report about Their Toxic Trail and Safer Alternatives: Chemical Treatment of Wood," n.d., 4. http://www.beyond pesticides.org/wood/pubs/poisonpoles/wood.html.

4. Testing was performed by the University of West Florida's Center for Diagnostics and Bioremediation, and the analysis for dioxin was performed by Analytical Perspectives in Wilmington, N.C. Wilma Subra, personal communication, October 16, 2007. Subra notes that this study is forthcoming and has been accepted for publication. For further information see CDB's Web site on this dioxin study at https://nautical.uwf.edu/webpresence/utils/search.cfm.

5. Steve Curwood and John Rudolph, "Reaching Beyond 'Mount Dioxin,'" Living on Earth, World Media Foundation, April 19, 1996, 4–5. http://www.loe.org/shows/shows.htm?programID=96-P13-00016.

6. U.S. Environmental Protection Agency, Region 4 Superfund, "Escambia Wood Treating Update," June 15, 2007, 1. See also U.S. Environmental Protection Agency Superfund Proposed Plan Fact Sheet Escambia Wood Treating Company Superfund Site Operable Unit 2–Groundwater, June 2008. www.etccleanup .org/.../ETC%20OU2%20Proposed%20Plan%20Final.pdf.

7. Williams, "Environmental Justice," 1.

8. Benjamin R. Chavis and Charles Lee, *Toxic Waste and Race in the United States: A National Report on the Racial and Socioeconomic Characteristics of Communities Surrounding Hazardous Waste Sites* (New York: United Church of Christ Commission for Racial Justice, 1987), cited in Daniel R. Faber and Eric J. Krieg, "Unequal Exposure to Ecological Hazards: Environmental Injustices in the Commonwealth of Massachusetts," *Environmental Health Perspectives* 110, Suppl. 2 (2002): 279.

9. Nekeita Taylor-Hunley, "Mount Dioxin: Pensacola's Big Dirty Secret," *ECorsair*, September 12, 2007, 2.

10. Wilma Subra, "Clarinda Triangle: Alternative 2 Capping/Containment," personal communication, June 1, 2007.

11. Scott Streeter, "EPA Evaluates Replacing Tarp Covering 'Mount Dioxin:' Stability during Hurricane Spurs Cost Evaluation," *Pensacola News Journal*, April 1, 2003, 1.

12. Edmond Tsang and John Reis, "Mt. Dioxin, National Academy of Engineering, 4, http://ww.onlineethics.org/CMS/profpractice/ppcases/numericalprob/ mtdioxin.aspx.

13. Frances Dunham, "Relocation Revisited: Pensacola Neighborhoods Begin Moving from Contaminated Homes," *Harbinger*, Mobile, Alabama, April 14, 1998, 3.

14. Streeter, "EPA Evaluates," 2.

15. Brian Cabell, "Neighborhood Finally Moving away from Mount Dioxin," CNN, October 18, 1996, 3.

16. Tsang and Reis, "Mt. Dioxin," 8.

17. Citizens Against Toxic Exposure, "The Pensacola Struggle for Justice Continues: Historic Relocation Still Has Serious Flaws," n.d., 3. For access to this report contact CATE at http://www.cate.ws/.

18. Joel S. Hirschhorn, "Two Superfund Environmental Justice Case Studies" (Wheaton, MD: Hirschhorn & Associates,), 3., n.d.

19. Edmund Tsang and Jon C. Reis, "Engineering Ethics Cases, Mechanical Engineering Case 8, Mount Dioxin," National Science Foundation and Bovay Fund Workshop, Texas A&M University, August 14–18, 1995, 3.

20. Frances Dunham, Citizens against Toxic Exposure, "Environmental Justice Against All Odds," n.d., 2.

21. Lois Marie Gibbs et al., *Dying from Dioxin: A Citizen's Guide to Reclaiming Our Health and Rebuilding Democracy* (Boston: South End Press, 1995), 361.

22. CATE, "The Pensacola Struggle for Justice Continues," 3.

23. Dunham, "Environmental Justice Against All Odds," 5.

24. U.S. Environmental Protection Agency, "Escambia Wood Update," 1.

25. CATE, "The Pensacola Struggle for Justice Continues," 6.

26. Dunham, "Relocation Revisited," 2.

27. Ibid. See also an EPA roundtable discussion of fenceline relocation issues: http://www.epa.gov/compliance/resources/publications/ej/nejac/nejacmtg/roundtable-relocation-0596.pdf.

28. CATE, "The Pensacola Struggle for Justice Continues," 7.

29. Office of the Inspector General, U.S. Environmental Protection Agency, "Replacement Housing at the Austin Avenue Radiation Site," #810009, executive summary, last updated September 28, 2007, 1.

30. Francis Dunham, "Superfund Relocation Racism: A White Community in Pennsylvania; An African-American Community in Florida," *Harbinger*, Mobile, Alabama, May 12, 1998, 1.

31. U.S. Environmental Protection Agency, Office of Inspector General, Audit Report, "Replacement Housing at the Austin Avenue Radiation Site," Executive Summary, March 30, 1998.

32. Office of the Inspector General, "Replacement Housing at the Austin Avenue Radiation Site," 2.

33. Hirschhorn, "Two Superfund Environmental Justice Case Studies," 5.

34. Dunham, "Environmental Justice," 2.

35. Wilma Subra, "Surface Soils Exceeding Florida Residential Cleanup Values," letter to Margaret Williams, president, CATE, December 28, 2004.

36. U.S. Environmental Protection Agency, "Escambia Wood Update," 1.

37. Subra, "EPA Proposed Plan for Escambia Treating Company Superfund Site," 9.

38. Local governments can declare blighted areas or places without adequate affordable housing Community Redevelopment Areas and then use tax increment financing to redevelop them. The EPA also provides small "pilot" grants to clean up formerly industrialized sites that are contaminated. http://www.epa.gov/superfund/programs/reforms/reforms/2-4a.htm.

39. U.S. Environmental Protection Agency, "Superfund Proposed Plan Fact Sheet," 15–18, 21–22. L. Spence, "Race, Class, and Environmental Hazards: A Study of Socio-Economic Association with Hazardous Waste Generators and Treatment/Storage/Disposal Facilities in Massachusetts (master's thesis, Tufts University, 1995), cited in Faber and Krieg, "Unequal Exposure to Ecological Hazards," 279.

40. Subra, "EPA Proposed Plan for Escambia Treating Company Superfund Site," 3.

41. This "final" phase of the cleanup refers to the remediation of contaminated soils and does not address the unresolved problem posed by groundwater contaminated at this Superfund site.

42. Derek Piynick, "Superfund Cleanup Gets $2 Million: Some Unhappy with Mt. Dioxin Burial Plan," *Pensacola News Journal,* June 2, 2007.

43. Erin Gibson, "Superfund Study Wants You: Those Who Lived Close to Sites Eligible for Free Health Screening," *Pensacola News Journal.*

44. Dunham, "Environmental Justice Against All Odds," 2.

45. CATE, "The Pensacola Struggle for Justice Continues," 1.

46. Dunham, "Environmental Justice Against All Odds," 3.

47. Hirschhorn, "Two Superfund Environmental Justice Case Studies," 4.

48. Dunham, "Environmental Justice Against All Odds," 4.

49. Citizens Against Toxic Exposure, "Escambia Treating Company and Agrico Superfund Sites: Contaminants of Concern and Health Effects," n.d., 1.

50. Tsang and Reis, "Engineering Ethics Cases," 5.

51. Williams, "Environmental Justice," 4.

## Chapter 3

1. Beth O'Brien, "Industrial Upset Pollution: Who Pays the Price? An Analysis of the Health and Financial Impacts of Unpermitted Industrial Emissions," Public Citizen, August 2005, 35.

2. Ibid.

3. Wilma Subra, personal communication with the author, January 19, 2007.

4. O'Brien, "Industrial Upset," 36.

5. Neil Carman, personal communication, January 14, 2007.

6. J. Levy, J. Spangler, D. Hinka, and D. Sullivan, "Estimated Public Health Impact of Criteria Air Pollutant Emissions from the Salem Harbor and Brayton Point Power Plants" (Cambridge, MA: Harvard School of Public Health and Sullivant Environmental Consulting, May 2000), cited in Daniel E. Faber and Eric J. Krieg, "Unequal Exposure to Ecological Hazards: Environmental Injustices in the Commonwealth of Massachusetts," *Environmental Health Perspectives* 110, Suppl. 2 (2002): 285.

7. Paul Ryder, "Good Neighbor Campaign Handbook: How to Win," *iUniverse,* Lincoln, Neb., 2006. http://www.ohiocitizen.org/about/finalinside.pdf.

8. Hilton Kelley, testimony to U.S. Senate Environment and Public Works and Judiciary Committees, July 16, 2002. http://judiciary.senate.gov/hearings/testimony.cfm?id=316&wit_id=733.

9. Donavan Webster and Michael Scherer, "No Clear Skies," *Mother Jones,* September/October 2003.

10. Kelley, testimony.

11. Sanford Lewis, "The Safe Hometowns Guide," Safe Home Towns Initiative, n.d. See http://www.citizen.org/documents/Chemical.pdf.

12. Michael May, "Port Arthur Blues: A Native Son Returns to Revitalize His Pollution-Plagued Neighborhood," *Texas Observer*, March 1, 2002.

13. Vicki Wolf, "Hilton Kelley: Standing Up for the West Side," Citizens League for Environmental Action Now (CLEAN), Houston, Texas, n.d. See http://www.cleanhouston.org/heros/kelley.htm.

14. O'Brien, "Industrial Upset," 68.

15. Eric V. Schaeffer and Huma Ahmed, "Accidents Will Happen: Pollution from Plant Malfunctions, Startups, and Shutdowns in Port Arthur, Texas," a report published by the Environmental Integrity Project, October 17, 2002, 4. http://www.environmentalintegrity.org/pub74.cfm. Copies of the report are available by writing EIP, 1875 Connecticut Avenue, NW, Suite 610, Washington, DC 20009. According to Schaeffer and Ahmed, January 21, 2002, was a particularly bad day for air quality. On that day BASF's-Altofina's ethylene plant in Port Arthur released 65 tons (130,805 pounds) of volatile organic compounds. Included in that release were xylene, which affects the brain and causes headaches, memory loss, dizziness, and confusion, as well as skin, eye, nose, and throat irritation and difficulty breathing; toluene, which affects the nervous system and can cause fatigue, confusion, weakness or kidney problems; and hexane, which causes muscle weakness and numbness.

16. Schaeffer and Ahmed, "Accidents Will Happen," 18.

17. Ibid., 12, 13.

18. O'Brien, "Industrial Upset," 32.

19. Ibid., 57.

20. Neil Carman, "Port Arthur–Beaumont Benzene Emissions, 1997 TRI Data," personal communication, January 13, 2007.

21. O'Brien, "Industrial Upset," 50.

22. Ibid., 72.

23. CLEAN, "Hilton Kelley."

24. Neil Carman, "Sierra Club Files Suit to Speed Up Beaumont-Port Arthur Clean Air Plan," Lone Star Sierran, Newsletter of the Sierra Club Lone Star Chapter, July 17, 2001. See http://lonestar.sierraclub.org/press/archive2001.asp.

25. People Against Contaminated Environments et al., "Beaumont-Port Arthur SIP Comment Letter," January 26, 2001, 10.

26. Neil Carman, "Background: Beaumont–Port Arthur Air Pollution Problems," personal communication, January 16, 2007.

27. Under the Clean Air Act, the U.S. Environmental Protection Agency gives different designations for the nonattainment of eight-hour ground-level ozone pollution. Areas that are judged to be in "nonattainment" of the National Ambient Air Quality Standards for smog are described as "basic," "moderate," "serious," "severe," or "extreme." Each of these levels triggers a different set of requirements about what measures industrial polluters must take to reach attainment in a reasonable period of time. By lowering the designation from "severe" to "serious," the measures that needed to be taken by industries emitting pollutants

that cause ground-level ozone contamination were less onerous. Nevertheless, the overall deal with refineries required them to make significant investments in new pollution controls.

28. "Clean Air Elusive in Beaumont/ Port Arthur, Texas," *Environmental News Service,* April 2, 2004.

29. CLEAN, "Hilton Kelley."

30. Global Community Monitor, CIDA, Refinery Reform Campaign, "Agreement Reached with Shell on Port Arthur Refinery Expansion," a press release published by Global Community Monitor, July 26, 2006. http://www.gcmonitor .org/article.php?id=405.

31. Ibid.

## Chapter 4

1. M. Harvey Weil, "The History of the Port of Corpus Christi 1926–2001," April, 1986. Website: Port of Corpus Christi. http://www.portofcorpuschristi .com/CHistory.html.

2. Public Citizen, "State Ignores 'Upset' Air Emissions by Industrial Plants At Tremendous Costs to Texas Taxpayers," *Human Health,* August 2, 2005. http://www.citizen.org/pressroom/release.cfm?ID=2011.

3. Anton Caputo, "Citgo Pollution Charges in Corpus Highlight Problems with the Law," *San Antonio Express News,* September 30, 2006, 1.

4. *Criteria air contaminants* are regulated by the U.S. Environmental Protection Agency under the National Ambient Air Quality Standards. The EPA has standards for six outdoor criteria air contaminants including: ozone, particulate matter, carbon monoxide, sulfur dioxide, nitrogen oxides, and lead. http://www.epa .gov/air/urbanair/.

5. Environmental Defense, "Chemical Profiles: Benzene," *Scorecard.* http://www .scorecard.org/chemical-profiles/summary.tcl?edf_substance_id=71% 2d43%2d2#hazards. Cited in "Industrial Upset Pollution: Who Pays the Price: An Analysis of the Health and Financial Impacts of Unpermitted Industrial Emissions," Public Citizen, August, 2005, 30. http://www.citizen.org/documents/ 08.01.05%20Industrial%20upsets%20report.pdf.

6. National Institute for Occupational Safety and Health, "NIOSH Pocket Guide to Chemical Hazards," http://www.cdc.gov/niosh/npg/pgintrod.html, cited in "Industrial Upset Pollution: Who Pays the Price." Ibid.

7. Texas Public Employees for Environmental Responsibility, PEER, "Corpus Christi's Refinery Row," *Toxic Texas,* n.d., 2. http://www.txpeer.org/toxic tour/corpus_christi.html.

8. K. C. Donnelly, Texas A&M School of Rural Public Health, Coastal Bend Health Education Center, Citizens for Environmental Justice, "Results of Blood and Urine Study in Hillcrest: Children and Adults Are Being Exposed to Volatile

Organic Compounds," November 7, 2008, 1. http://www.citgojustice.org/PDF/
biomonitoring%20study%20press%20release%20nov%2008.pdf.

9. Dan Kelley, "Study Shows High Level of Benzene," Global Community Monitor, November 8, 2008. http://www.gcmonitor.org/article.php?id=820.

10. Suzie Canales, "Corpus Christi Texas. Criminal Injustice in an All-American City: Toxic Crimes, Race-Zoning, and Oil Industry Cover-Up," Citizens for Environmental Justice and Global Community Monitor, 10. http://www.citgojustice.org/GCMcorpus.pdf.

11. Caputo, "Citgo Pollution."

12. Ibid.

13. 2003 Contaminant Summary Report for Valero Refining Company NE0112G, cited in "Industrial Upset Pollution: Who Pays the Price," 15.

14. Suzie Canales, "SEPs: Supplemental Environmental Projects: The Most Affected Communities Are Not Receiving Satisfactory Benefits," Public Citizen and Refinery Reform Campaign, June 2006, 9. http://www.citgojustice.org/PDF/SEPs%20Report-SCanales-061906.pdf.

## Chapter 5

1. Mike Rutledge and Dan Klepal, "Neighbors Worry about Plant: Odors, Pollution Raise Health Questions," *Cincinnati Enquirer*, December 11, 2004.

2. Dan Klepal, "Addyston Residents Monitoring Lanxess," *Cincinnati Enquirer*, March 23, 2005.

3. Dan Klepal and Mike Rutledge, "Gas Leaked from Plant: Addyston Plastics Firm Released 1,200 Lbs. during Festival," *Cincinnati Enquirer*, December 4, 2004.

4. Dan Klepal, "Stop Gas Releases, Firm Told," *Cincinnati Enquirer*, April 9, 2005.

5. Klepal and Rutledge, "Gas Leaked from Plant."

6. Rutledge and Klepal, "Neighbors Worry about Plant."

7. Dan Klepal, "Lanxess Plant Released More Gas: Company Cites Reactor Problem," *Cincinnati Enquirer*, January 27, 2005.

8. Dan Klepal, "Cloud Had 750 Pounds of Toxin: Addyston Company Reviews after Third Gas Release," *Cincinnati Enquirer*, March 11, 2005, B1.

9. Jay Richey, "Lanxess Working to Address Community Concerns," *Western Hills Press,* April 13, 2005.

10. Dan Klepal, "Study: Lanxess Releases Not a Risk," *Cincinnati Enquirer*, May 24, 2005.

11. Dan Klepal, "Release of Gas at Lanxess Plant in Addyston Is Third since October," *Cincinnati Enquirer*, February 26, 2005.

12. Ruth Breech, "Lanxess Can Be a Good Neighbor, Work with Community," *Western Hills Press,* April 6, 2005, A4.

13. Hagit Limor, "Chemical Plant in Tri-State Neighborhood," 9News, WCPO, Cincinnati, Ohio, May 5, 2005.

14. Ibid., 5.

15. Ibid., 3–4.

16. Gregory Korte, "City Can't Control Chemical Odors: Clean Air Rule Not Enforceable," *Cincinnati Enquirer*, April 26, 2005.

17. Dan Klepal, "Plant Releases Volatile Gases," *Cincinnati Enquirer*, April 1, 2005.

18. Dan Klepal, "Addyston Cancer Rates to Be Studied: State Cluster Review Comes after Lanxess Finding," *Cincinnati Enquirer*, December 9, 2005.

19. Peggy O'Farrell, "Addyston Cancers Studied," *Cincinnati Enquirer*, June 5, 2006, B1.

20. Ibid.

21. Dan Klepal and Denise Smith Amos, "Cancer Risk Closes Addyston School," *Cincinnati Enquirer*, December 6, 2005.

22. Peggy O'Farrell, "Addyston Cancer Rate 'Troubling,'" *Cincinnati Enquirer*, May 26, 2006, A1.

23. Peggy O'Farrell, "Lanxess Says Addyston Cancer Study Flawed, Chemical Company Denies It Is the Cause of High Incidence," *Cincinnati Enquirer*, June 17, 2006, B1.

24. Dan Klepal, "Air Checked next to Lanxess," *Cincinnati Enquirer*, May 6, 2005.

25. Dan Klepal, "Plastics Maker Questions Monitor Test," *Cincinnati Enquirer*, June 6, 2005.

26. Dan Klepal, "Study: Lanxess Releases Not a Risk," *Cincinnati Enquirer*, May 24, 2005.

27. Dan Klepal, "Bucket Brigades Celebrate: Efforts to Fight Air Pollution Began 10 Years Ago," *Cincinnati Enquirer*, October 16, 2005.

28. Jay Warren, "School near Controversial Chemical Plant Closes," 9News, WCPO Cincinnati, Ohio, December 5, 2005.

29. Spenser Hunt, "Polluted Air Prompts Town to Close School," *Columbus Post Dispatch*, December 6, 2005.

30. Klepal and Amos, "Cancer Risk Closes Addyston School," *Cincinnati Enquirer*, December 6, 2005.

31. Spencer Hunte, "Tough Times in Addyston," *Columbus Post Dispatch*, December 7, 2005.

32. Hagit Limor, "Lanxess Cited by EPA," 9News, WCPO, Cincinnati, June 14, 2006, 1–2.

## Chapter 6

1. Spencer Hunt, "Blue Smoke, Tainted Water: Worrisome Neighbors: Many Ohio River Residents Wonder How Factories, Plants Affect Their Health," *Columbus Dispatch*, December 4, 2005.

2. Ohio Citizen Action, "Dear Stink Diary," October 30, 2006. http://www .ohiocitizen.org/campaigns/eramet/diary.html.

3. Ibid., 4.

4. Neighbors for Clean Air and Ohio Citizen Action, "Good Neighbor Campaign: Citizen Audit of Eramet Marietta," June 1, 2006. http://www.ohiocitizen .org/campaigns/eramet/Citizen%20Audit%20final%20master.pdf.

5. Ibid., 4.

6. Hunt, "Blue Smoke, Tainted Water," 5.

7. David Pace, "Mid Ohio Valley Factories Linked to Dangerous Air," Associated Press, December 15, 2005. http://hosted.ap.org/specials/interactives/ archive/pollution/part2.html.

8. Ohio Citizen Action, "Eramet News from March 2000 to June 2006." http://www.ohiocitizen.org/campaigns/eramet/erametb.htm.

9. Sam Shawver, "Study of Local Air Continues," *Marietta Times*, September 13, 2006. http://www.ohiocitizen.org/campaigns/eramet/091306.htm.

10. Dennis Kuhl, letter to the Editor, *Marietta Times*, December 26, 2005.

11. Daniel R. Faber and Eric J. Krieg, "Unequal Exposure to Ecological Hazards: Environmental Injustices in the Commonwealth of Massachusetts," *Environmental Health Perspectives* 110, Suppl. 2 (2002): 279.

12. Lesley Kuhl, "Parents Want Tighter Controls on Pollution," *Marietta Times*, October 14, 2006.

13. Lesley Kuhl, "Viewpoint: Voice Concerns about Local Air Pollution to Officials Who Can Make a Difference," *Marietta Times*, March 18, 2006, Ohio Citizen Action, "Eramet News from March 2000 to June 2006," http://www .ohiocitizen.org/campaigns/eramet/erametb.htm.

14. David Pace, "Parents Ask How Air Pollution Affects Kids," Associated Press, December 14, 2005. http://hosted.ap.org/specials/interactives/archive/ pollution/part2.html.

15. Ibid.

16. Brad Bauer, "Airborne Manganese Levels Assessed: Study Focusing on Eramet Area Viewed as Tool for Scientists," *Marietta Times*, July 25, 2007, 2; Pace, "Parents Ask How Air Pollution Affects Kids."

17. Jeff McKinney, "Release of 'Citizen Audit' of Eramet Marietta," *For the Record*, June 13, 2006, 5. http://www.ohiocitizen.org/campaigns/eramet/ftrecord3.pdf.

18. Tom Lotshaw, "University of Cincinnati Researcher Visits Marietta to Give Update on Children's Health Survey," *Marietta Register*, October 22, 2008.

19. University of Cincinnati Academic Health Center, "Researchers Assess Exposure to Metal Emissions in Marietta," October 24, 2006.

20. Connie Cartmell, "UC Study to Check Manganese," *Marietta Times*, July 4, 2006.

21. Callie Lyons, "Researcher Tries to Get a Handle on Potentially Hazardous Plant Emissions," *Athens News*, January 7, 2008.

22. Ibid., 4.

23. Bauer, "Airborne Manganese Levels," 1.

24. Tom Lotshaw, "Eramet Announces Vision of Potential Facility Upgrades," *Marietta Register*, September 3, 2008.

25. Pace, "Parents Ask How Air Pollution Affects Kids," 2.

26. Ellyn Burnes, *Parksburg News*, April 10, 2006, cited in Ohio Citizen Action, "Eramet News from March 2000 to June 2006," 10, http://www.ohiocitizen.org/campaigns/eramet/erametb.html.

27. Pace, "Parents Ask How Air Pollution Affects Kids," 2.

28. Angel Oster, Ohio Citizen Action, "Eramet Neighbors Speak Out," DVD.

29. Caroline Beidler, "Community Meeting on Eramet," February 7, 2006, in "Eramet News March 200 to June 2006," Environmental Campaigns, Ohio Citizen Action. http://www.ohiocitizen.org/campaigns/eramet/erametb.htm.

30. Ohio Citizen Action, Environmental Campaigns, "Q. Where Is the Most Dangerous Air Pollution in the Country? A. Marietta, Ohio," n.d. http://www.ohiocitizen.org/campaigns/eramet/problem.htm. This report cites "Bathed in toxins, with unanswer questions," Associated Press, *St Petersburg Times*, December 18, 2005.

31. Sandy Buchanan, "Ohio Citizen Action Good Neighbor Campaigns 1998–2008, 6. http://www.ohiocitizen.org/campaigns/good/gnc.bkgd.html.

32. Neighbors for Clean Air and Ohio Citizen Action, "Good Neighbor Campaign: Citizen Audit of Eramet Marietta," June 1, 2006, 8. http://www.ohiocitizen.org/campaigns/eramet/Citizen%20Audit%20final%20master.pdf.

33. Ohio Citizen Action, "Good Neighbor Campaign Victory," http://www.ohiocitizen.org/campaigns/eramet/erametg.html.

34. McKinney, "Release of 'Citizen Audit,'" 6. http://www.ohiocitizen.org/campaigns/eramet/ftrecord3.pdf.

35. Ohio Citizen Action, "Eramet News from March 2000 to June 2006." http://www.ohiocitizen.org/campaigns/eramet/erametb.html.

36. Oster, "Eramet Neighbors Speak Out."

37. Vicki Dils, letter to the editor, *Parksburg News and Sentinel*, April 30, 2006.

38. McKinney, "Release of 'Citizen Audit,'" 1. http://www.ohiocitizen.org/campaigns/eramet/ftrecord3.pdf.

39. Ohio Citizen Action, "Eramet News from March 2000 to June 2006," 13, www.ohiocitizen.org/campaigns/eramet/erametb.html.

40. Joseph Koncelik, director, Ohio EPA, letter to Ruth Breach, July 20, 2006.

41. Tom Lotshaw, "Eramet Announces Plant Upgrade," *Marietta Register*, August 30, 2008.

## Chapter 7

1. Robert Cilek et al., "Tallevast Community Preliminary Contamination Assessment Report," (PCAR/FDEP), SIS Report no. 2004–01, July 2004, 2.

2. Donna Wright, "Tallevast Plume History," *Bradenton Herald*, June 4, 2006.

3. One of the best sources of the history of the contamination in Tallevast is contained in a letter to Winston Smith, director, Waste Management Division, U.S. EPA in Atlanta, from Michael W. Sole, director, Florida Division of Waste Management, June 29, 2004. www.dep.state.fl.us/secretary/news/2004/tal/epa_letter .pdf.

4. Cilek et al., "Tallevast Community Preliminary Assessment Report," 3.

5. Wilma Subra, "Tallevast Timeline of Contamination," n.d. This report was written by Subra for the *Bradenton Herald* and at the request of community members.

6. Cilek et al., "Tallevast Community Preliminary Assessment Report," 4.

7. Ibid., 13.

8. Ibid., 15.

9. Wright, "New Tallevast Tests Ordered," *Bradenton Herald*, February 2, 2006.

10. Ibid.

11. Donna Wright, "Scientist Warns of Tallevast Dangers," *Bradenton Herald*, September 18, 2005.

12. "Opinion," *Bradenton Herald*, April 16, 2006.

13. Duane Marsteller, "Tallevast Plume 50 Percent Larger," *Bradenton Herald*, April 28, 2006.

14. Donna Wright, "Lockheed Set to Start Cleaning Tallevast Contamination Site: Toxins to Be Pumped out of Groundwater," December 21, 2005, *Bradenton Herald*.

15. "Lockheed Martin and Others Sued for Groundwater and Soil Contamination in Florida; Claim Alleges that Corporations Knew Toxins Migrated Off-Site but Failed to Inform Residents," Business Wire, September 6, 2005. http://find articles.com/p/articles/mi_m0EIN/is_2005_Sept_6/ai_n15343705/.

16. Richard Girard, "The Weapons Manufacturer That Does It All: A Profile of an Arms Giant, Lockheed Martin" (Ottawa: Polaris Institute, November 2005), 1, 8.

17. Ibid., 30.

18. Ibid., 29.

19. Ibid., 34.

20. Ibid., 35.

21. Ibid., 36.

## Chapter 8

1. Tanji Patton, "Toxic Triangle? Is Contamination Connected to Liver Cancer Cases?" News 4 WOAI, November 15, 2006.

2. Anton Caputo and Jerry Needham, "Kelly Cleanup May Get a Whole Lot Costlier," *San Antonio Express News*, July 31, 2006.

3. Ralph Vartabedian, "Cancer Stalks a Toxic Triangle," *Los Angeles Times*, March 30, 2006, 1.

4. Ibid.

5. Gilbert Garcia, "Containment Policy: For Residents of Kelly's Toxic Triangle, Answers Are Hard to Come By," *San Antonio Current*, July 28–July 4, 2006.

6. Ibid.

7. Gilbert Garcia, "For ailing residents of Kelly's toxic triangle, answers are hard to come by," *San Antonio Current*, June 28, 2006.

8. Southwest Public Workers' Union, "North Kelly Gardens Community Health Survey near Kelly AFB" (rev. ed.), October 23, 1997, cited in "Kelly Air Force Base: San Antonio's Dumping Ground," *Toxic Texas*, Public Employees for Environmental Responsibility, 1. n.d. http://www.txpeer.org/toxictour/kelly.html#top.

9. Anton Caputo and Jerry Needham, "Kelly Cleanup may get a whole lot costlier," *San Antonio Express News*, July 29, 2006.

10. Interview with Yolanda Johnson cited in Public Employees for Environmental Responsibility "Toxic Texas: Kelly Air Force Base: San Antonio's Dumping Ground,," 2.

11. Public Employees for Environmental Responsibility, "Toxic Texas: Kelly Air Force Base: San Antonio's Dumping Ground," 2.

12. Caputo and Needham, "Kelly Cleanup May Get a Whole Lot Costlier."

13. Southwest Workers Union, Committee for Environmental Justice Action, "Fact Sheet: Kelly AFB Struggle." n.d. http://www.swunion.org.

14. Southwest Workers Union, Committee for Environmental Justice Action, "Organizing Environmental Justice Military Toxics," n.d., 2.

15. Roddy Stinson, "Kelly Contamination Investigation Pays Off," *San Antonio Express News*, March 22, 1998, 3A.

16. "Community Groups File Civil Rights Complaint Pertaining to Activities at Kelly Air Force Base," press release, May 4, 1999, 2, cited in Public Employees for Environmental Responsibility, "Toxic Texas: Kelly Air Force Base: San Antonio's Dumping Ground," 4.

17. Caputo and Needham, "Kelly Cleanup May Get a Whole Lot Costlier."

18. Jerry Needham, "Activists Urge Cleanup at U.S. Bases Worldwide," *San Antonio Express News*, July 14, 2006.

19. Patton, "Toxic Triangle?"

20. Permeable reactive barriers use a wide variety of materials to trap or render harmless dissolved groundwater contaminants.

21. Garcia, "Containment Policy."

22. Southwest Workers Union, Committee for Environmental Justice Action, "Fact Sheet: Kelly AFB Struggle."

23. Vartabedian, "Cancer Stalks a Toxic Triangle."

24. Ibid.

25. Ibid.

26. Patton, "Toxic Triangle?"

27. Kelli Dailey, "Work Related (Or: How Many Lives Can a Carcinogen Touch?)" *San Antonio Current*, September 13, 2006.

28. Sunaura Taylor and Astra Taylor, "Military Waste in Our Drinking Water," EnviroHealth, *Alternet*, August 4, 2006. http://www.alternet.org/environment/39723.

29. ALS Association, "Kelly Air Force Base Update, January 22, 2002." The ALS Association, a nonprofit, prepares white papers on various topics. The white paper listed in this note and in notes 30, 31, and 32 include comments by three of their experts on the JOEM article listed below. The ALS Association experts include: Carmel Armon, MD, MHS; Merit Cudkowicz, MD; and Jean Brender, Ph.D. Most of these white papers are undated. "Cause-Specific Mortality among Kelly Air Force Base Civilian Employees, 1981–2001," *Journal of Occupational and Environmental Medicine* 44 (November 2002): 989–996. Diane J. Mundt, PhD; Linda D. Dell, MS; Rose S. Luippold, MS; Sandra I. Sulsky, PhD; Anne Skillings, MS; Rachel Gross, MS; Kenneth L. Cox, MD, MPH; Kenneth A. Mundt, PhD.

30. ALS Association, "Kelly Air Force Base Study Results: Clarifications and Corrections," n.d.

31. ALS Association, "Kelly Air Force Base Study Results: Dr. Carmel Armon's Comments," n.d.

32. ALS Association, "ALS in the Military: Unexpected Consequence of Military Service," May 11, 2005, 5, 9.

## Chapter 9

1. Martin Kuz, "What Lies Beneath," *San Francisco Weekly*, December 20–26, 2006, 18.

2. The EPA requires that hazardous wastes be disposed of in specialized facilities. The agency has five categories of hazardous waste disposal sites. The most dangerous toxic wastes, which present an imminent threat to public health or the environment, must be placed in class 1 hazardous waste disposal sites.

3. Melissa McMillan, "Midway Village Housing Complex: A Struggle for Environmental Justice," www.greenaction.org/midway/.

4. Wilma Subra, "Review of the 2001 Investigation and Cleanup of Midway Village Residential Complex in Daly City, California," review draft, February 13, 2006. This report, authored by Wilma Subra for the Midway Village Review Committee, was sent to David Siegel, Chief, Integrated Risk Assessment Branch, Office of Environmental Health Hazard Assessment, and Charles Salocks, staff toxicologist in the same office.

5. Agency for Toxic Substances and Disease Registry, Division of Health Assessment and Consultation, "Health Consultation: Midway Village Site, Daly City, San Mateo County, California," April 8, 1999, 2.

6. McMillan, "Midway Village Housing Complex."

7. Kuz, "What Lies Beneath," 18.

8. Ibid., 20.

9. Ibid., 22.

10. These statistics were compiled from two sources: Agency for Toxic Substances and Disease Registry, Division of Health Assessment and Consultation, "Health Consultation: Midway Village Site, Daly City, San Mateo County, California," 3; Subra "Review of the 2001 Investigation and Cleanup of the Midway Village Complex in Daly City California."

11. Agency for Toxic Substances and Disease Registry, "Health Consultation: Midway Village Site, Daly City, San Mateo County, California," 5.

12. Subra, "Comments on the Review of the 2001 Investigation and Cleanup of Midway Village Residential Complex in Daly City, California, Draft Review, February 2006," February 13, 2006.

13. Wilma Subra, "Midway Village and Bayshore Park Site—Review of Documents Provided as Background Information," January 2, 2006. Report of the Midway Village Review Committee submitted to David Siegel, Chief, Integrated Risk Assessment Branch, Office of Environmental Health Hazard Assessment, and Charles Salocks, staff toxicologist in the same office.

14. "Midway Village/PG&E Fact Sheet," n.d.

15. Rosemarie Bowler, "Health Survey of Midway Village Residents Exposed to PNAs," San Francisco State University, Department of Psychology, July 19, 1996, 44.

16. Ibid., 44–45.

17. McMillan, "Midway Village Housing Complex," 3.

18. Kelly Forman, "Field Guide for Altgeld Gardens," University of Wisconsin, Department of Geography, n.d., http://www.geology.wisc.edu/~wang/EJBaldwin/PCR/Case%20Study%20One.doc.

## Chapter 10

1. Mimi Hogan, Sandra Christopherson, and Anne Rothe, "Formerly Used Defense Sites in the Norton Sound Region: Location, History of Use, Contaminants Present, and Status of Cleanup Efforts," Alaska Community Action on Toxics, July 1, 2006, 3.

2. Ibid., 13.

3. Alaska Community Action on Toxics, "Body Burden Testing of Residents on St. Lawrence Island," cited in *Community Monitoring Handbook*, Alaska Case Study, http://chemicalbodyburden.org/hb_cs_alaska.htm.

4. Hogan, Christopherson, and Rothe, "Formerly Used Defense Sites," 14.

5. The estimated size of the diesel spill is unclear, claims Carey Cossaboom of the Army Corps of Engineers. It may be anywhere from 30,000 to 220,000 gallons, he says.

6. Diana Haecker, "Cleanups of FUDS Continue in Nome and Region," *Nome Nugget*, August 9, 2007, 5.

7. Hogan, Christopherson, and Rothe, "Formerly Used Defense Sites."

8. Ibid., 4.

9. Bill Sherwonit, "Tracking Toxics," *Orion*, July–August 2003, 64.

10. Hogan, Christopherson, and Rothe, "Formerly Used Defense Sites," 14.

11. *Diesel range organics* (DRO) is a term that refers to petroleum substances with a range of carbon atoms from C10 to C25.

12. U.S. Army Corps of Engineers, "Fact Sheet: Proposed Plan Northeast Cape Air Force Station, St. Lawrence Island," FUDS#F10AK096903, July 2007, 1–42.

13. ACAT, "Body Burden Testing of Residents on St. Lawrence Island," 3.

14. Ibid., 2.

15. Pam Miller, "Military Sites in Alaska," Alaska Community Action on Toxics, 3, n.d. http://www.akaction.org/overview_Military_Sites_in_Alaska_Impacts_to_Environment_and_Communities.htm.

16. Ibid., 2.

17. Lori Verbrugge et al., "PCB Blood Test Results from St. Lawrence Island: Recommendations for Consumption of Traditional Foods," *Bulletin*, February 6, 2003, 1–5. This was a statement from the Alaska Division of Public Health, State of Alaska Epidemiology.

18. Sherwonit, "Tracking Toxics," 64.

19. Vi Waghiyi, coordinator, Alaska Community Action on Toxics, St. Lawrence Island Environmental Justice Project, testimony at the National Environmental Justice Advisory Council, April 14, 2004, http://www.akaction.org/Testimony_of_Viola_Waghiyi.htm.

20. Carey Cossaboom of the U.S. Army Corps of Engineers notes that the soils in Gambell have been tested and no PCBs have been detected.

21. Jeanette Lee, "$1 B Cleanup of Alaska Bases Drags On," Associated Press, August 24, 2007. "Money Shrinks for $1 Billion Cleanup of Deserted Military Sites near Alaska Villages," USAToday.com, August 24, 2007.

22. Army Corps of Engineers, "Fact Sheet. Decision Document: Gambell Formerly Used Defense Site, St. Lawrence Island, Alaska," June 2005, 4.

23. Government Accountability Office, "Lessons Learned from the Cleanup of Formerly Used Defense and Military Munitions Sites," June 10, 2009. http://www.gao.gov/new.items/d09779t.pdf.

24. ACAT, "Body Burden Testing of Residents on St. Lawrence Island," 3.

25. Cossaboom disagrees with this account and claims that the debris that would have been buried in the monofill was not contaminated.

26. Carey Cossaboom, personal communication, October 26, 2007.

27. Cossaboom disagrees with this analysis, arguing that the EPA never listed the Northeast Cape as an NPL Superfund site and that in 2002 the agency investigated the Army Corps cleanup process there and found that it was "proceeding in a manner consistent with EPA expectations for hazardous waste sites."

28. Waghiyi, testimony, 3.

29. Charles Lee, "Environmental Justice: Building a Unified Vision of Health and the Environment," *Environmental Health Perspectives* 110, Suppl. 2 (2002): 143. Brownfields Showcase Communities are federally funded projects aimed at demonstrating the environmental and economic benefits of cleaning up idle or abandoned commercial properties where redevelopment was complicated by contamination problems.

## Chapter 11

1. Julie Leibach, "Black Mayonnaise," *Science Line*, January 24, 2007, 1.

2. Daphne Eviatar, "The Ooze: Ten Million Gallons of Toxic Gunk Trapped in the Brooklyn Aquifer Is Starting to Creep toward the Surface. How Scary Is That?" *New York Magazine*, June 3, 2007, 4.

3. Erik Baard, "East River Tributaries: Newtown Creek," *Waterwire.net*, Metropolitan Waterfront Alliance, December 11, 2006, 5.

4. Ibid., 2.

5. Leibach, "Black Mayonnaise, 1.

6. Ibid., 5.

7. Eviatar, "The Ooze," 5.

8. Leibach, "Black Mayonnaise," 2.

9. Matthew Leising, "Exxon Mobil's Biggest Oil Spill Is in Brooklyn, Not Alaska," *International Herald Tribune*, January 4, 2007.

10. Eviatar, "The Ooze," 6.

11. Ibid., 3.

12. Ibid., 6.

13. Nathan Duke, "Exxon Slows Creek Cleanup," *TimesLedger.Com*, March 15, 2007.

14. Eviatar, "The Ooze," 7.

15. Ibid.

16. Andrew M. Cuomo, Attorney General, State of New York, "Notice of Intent to Sue ExxonMobil Corporation and ExxonMobil Refining & Supply Company for Clean Water Act Violations in Greenpoint, Brooklyn, New York," February 8, 2007, 1, 2, 5.

17. Andrew M. Cuomo, "Attorney General Andrew Cuomo to Sue Exxon-Mobil in Landmark Case to Clean Up Catastrophic Greenpoint Oil Spill and Newtown Creek Contamination," Office of the New York State Attorney General, February 8, 2007, 1.

18. Ibid., 2.

19. Leibach, "Black Mayonnaise," 6.

20. Baard, "East River Tributaries," 1, 2.

**Chapter 12**

1. Harold Curan, *Fearful Crossing: The Central Overland Trail Through Nevada* (Reno, NV: Great Basin Press, 1982), 181.

2. Frank X. Mullen, Jr., "Scientists: Environment May Factor into Cancer Mystery," *Reno Gazette-Journal*, October 16, 2007, 2.

3. Frank X. Mullen Jr., "Fallon Families Battle Back," *Reno Gazette-Journal*, August 16, 2002, 1.

4. Victoria Pearson, "Professor Speaks about Leukemia Cluster Findings," *Lohontan Valley News*, May 10, 2007.

5. Renee Downing, "For the Kids: Since Government Agencies Often Ignore or Dismiss Disease Clusters, Parents and Scientists Are Taking Matters into Their Own Hands," *Tucson Weekly*, May 18, 2006, 2.

6. Craig Steinmaus, Meng Lu, Randall L. Todd, and Allan H. Smith, "Probable Estimates for the Unique Childhood Leukemia Cluster in Fallon, Nevada, and Risks Near Other U.S. Military Aviation Facilities," *Environmental Health Perspectives* 112, no. 6 (2004): 3.

7. Renee Downing, "Cancer Wars: An Abnormal Number of Kids in Sierra Vista Are Getting Leukemia. Why Does the Government Insist That It Is Just a Coincidence?" *Tucson Weekly*, February 12, 2004, 5, 6.

8. C. S. Rubin et al., "Investigating Childhood Leukemia in Churchill County, Nevada," *Environmental Health Perspectives* 155, no. 1 (January 2007): 151–157.

9. Amie S. Williams, *Fallon, Nevada: Deadly Oasis*. Documentary, 2003, Bal-Maiden Films, http://www.balmaidenfilms.com/featuredocs.html.

10. Bill Moyers, *Now with Bill Moyers*, May 10, 2002, printable edition, 1, 3, 4, 5.

11. Richard Lake, "Focus on Fallon: Cancer Cluster Confounds Experts. Frustrated Families Look Elsewhere for Answers," *Las Vegas Review Journal*, November 17, 2002, 11.

12. Ibid., 2.

13. Floyd Sands, "Time Line of Death," August 29, 2007, 2, 3. http://www falloncancercrisis.org.

14. Frank X. Mullen Jr., "Panel Says No More Environmental Testing Recommended in Fallon," *Reno Gazette-Journal*, February 23, 2004, 1.

15. Dick Russell, Sanford Lewis, and Brian Keating, "Inconclusive by Design: Waste, Fraud and Abuse in Federal Environmental Health Research" (Harvey, Louisiana and Boston, Mass.: Environmental Health Network, and National Toxics Campaign Fund, May, 1992).

16. Lake, "Focus on Fallon," 5, 6.

17. Alexandra C. Miller et al., "Neoplastic Transformations of Human Osteoblast Cells to the Tumorgenic Phenotype by Heavy Metal Tungsten Alloy Particles: Induction of Genotoxic Effects," *Carcinogenesis* 22, no. 1 (January 2001): 115–125.

18. "Why Is There So Much Leukemia in Fallon, NV?" *Angry Toxicologist*, May 15, 2007, 1.

19. "Elevated Tungsten in Trees Coincides with Nevada Leukemia Cluster," *Science Daily*, University of Arizona, May 1, 2007, 1.

20. Pearson, "Professor Speaks about Leukemia Cluster Findings."

21. Frank X. Mullen Jr., "Study: Fallon Firm Is Source of Metal Implicated in Cancer Cluster," *Reno Gazette-Journal*, November 17, 2004, 1.

22. Downing, "Cancer Wars," 6.

23. Ibid.

24. Frank X. Mullen Jr., "Arizona Leukemia Cases May Help Fallon Probe," *Reno Gazette-Journal*, August 26, 2002, 1.

25. Downing, "Cancer Wars," 1.

26. Frank X. Mullen, Jr., "No Pollution Controls in Tungsten Plant," *Reno Gazette-Journal*, February 5, 2003, 1.

27. Paul R. Sheppard, Paul Toepfer, Elaine Schumacher, Kent Rhodes, Gary Ridenour, and Mark L. Witten, "Morphological and Chemical Characteristics of Airborne Tungsten Particles of Fallon, Nevada," *Microscopy and Microanalysis* 13 (2007): 296–303.

28. Frank X. Mullen Jr., "Air Sampling Units Set Up in Fallon," *Reno Gazette-Journal*, March 19, 2004, 1.

29. Mullen, "Study," 1.

30. "Elevated Tungsten in Trees," 1.

31. Ibid.

32. Paul R. Sheppard, Robert Speakman, Gary Ridenour, and Mark L. Witten, "Temporal Variability of Tungsten and Cobalt in Fallon, Nevada," *Environmental Health Perspectives* 115, no. 5 (2005), Results section. See also Frank X. Mullen Jr., "Scientists Look for Volunteers to Take Air Sample Machines," *Reno Gazette-Journal*, March 29, 2004, 1.

33. Jill James, "Genetic/Metabolic Sensitivity to Oxidative Stress and Risk of ALL [Leukemia] in Fallon, Nevada," n.d. (Little Rock, AR: Children's Hospital Research Institute).

34. *Oxygen free radicals* are molecules that contain at least one unpaired electron. Oxidative stress is a condition of increased oxidant production in animal cells characterized by the release of free radicals and resulting in cellular degeneration.

35. Moyers, *Now with Bill Moyers*, 11.

36. Frank X Mullen Jr., "DNA May Hold Key to Cluster," *Reno Gazette-Journal*, May 13, 2001, 1.

37. Frank X. Mullen Jr., "Seeking Answers: Fallon Cancer Cluster," *Reno Gazette-Journal*, August 26, 2002, 2.

38. Lake, "Focus on Fallon," 10.

39. Jeffrey St. Clair, "Fallon's Fallen: Is the U.S. Navy Killing Children in Nevada?" Counterpunch.org, November 6, 2007, 2.

40. Agency for Toxic Substances and Disease Registry, "Public Health Assessment: Naval Air Station Fallon," July 18, 2003, 2. http://www.atsdr.cdc.gov/HAC/pha/fallon/nas_toc.html.

41. Lake, "Focus on Fallon," 10.

42. ATSDR, "Public Health Assessment," 4.

43. Williams, *Fallon, Nevada: Deadly Oasis*.

44. Mullen, "Seeking Answers," 1.

45. Agency for Toxic Substances and Disease Registry, "Health Consultation: Evaluation of Potential Exposures from Fallon JP-8 Fuel Pipeline," updated June 7, 2002.

46. Frank X Mullen Jr., "Radioactivity Found in Fallon Wells," *Reno Gazette-Journal*, April 18, 2001, 1.

47. Frank X. Mullen Jr., "Children of Cancer: Experts: Viral or Bacterial Infection Most Likely Cause," *Reno Gazette-Journal*, August 26, 2002, 2.

48. Gross alpha emitting radioactive elements include:

- Americium-241
- Plutonium-236
- Uranium-238
- Thorium-232
- Radium-226
- Radon-222
- Polonium-210

The health effects of alpha particles depend heavily upon how exposure takes place. External exposure (external to the body) is of far less concern than internal exposure, because alpha particles lack the energy to penetrate the outer dead layer of skin. However, if alpha emitters have been inhaled, ingested (swallowed) or absorbed into the blood stream, sensitive living tissue can be exposed to alpha radiation. The resulting biological damage increases the risk of cancer; in particular, alpha radiation is known to cause lung cancer in humans when alpha emitters are inhaled. The greatest exposures to alpha radiation for average citizens comes from the inhalation of radon and its decay products, several of which also emit potent alpha radiation.

U.S. Environmental Protection Agency, "Gross Alpha Radiation Fact Sheet: http://www.stoller-eser.com/FactSheet/alpha.htm.

49. Frank X. Mullen Jr., "Two Fallon Dairies on Hold," *Reno Gazette-Journal*, August 4, 2007, 1.

50. Mullen, "Scientists," 2.

51. Agency for Toxic Substances and Disease Registry, "Health Consultation: Fallon Leukemia Project," February 12, 2003, 7. http://www.atsdr.cdc .gov/HAC/PHA/fallonleukemia2/fln_p1.html.

52. Ibid., 11, 18.

53. Ibid., 19.

54. Mullen, "Children of Cancer," 2.

55. Ibid., 1.

56. Mullen, "Scientists," 2.

Conclusion

1. Stephen Sandweiss, "The Social Construction of Environmental Justice," in *Environmental Injustices, Political Struggles*, ed. David E. Camacho (Durham, NC: Duke University Press, 1998), 46.

2. This calculation is supported by U.S. EPA Toxic Release Inventory data from 1989, cited in Daniel R. Faber and Eric J. Krieg, "Unequal Exposure to Ecological Hazards: Environmental Injustices in the Commonwealth of Massachusetts," *Environmental Health Perspectives* 110, Suppl. 2 (2002): 281–282.

3. Faber and Krieg, "Unequal Exposure to Ecological Hazards," 282.

4. Benjamin F. Chavis, "Environmental Justice Is Social Justice," *Los Angeles Times*, January 19, 1993, B7.

5. Robert D. Bullard, "Overcoming Racism in Environmental Decision-Making," *Environment*, May 1994, 11.

6. Cited in Claire Hassler, "The Purpose of the Environmental Justice Act: I Have a (Green) Dream," *University of Puget Sound Law Review* 17, no. 2 (1994): 417–418.

7. David Pace, "More Blacks Live with Pollution," Associated Press, December 13, 2005, 5. http://hosted.ap.org/specials/interactives/archive/pollution/ part1.html.

8. Sandweiss, "The Social Construction of Environmental Justice," 51.

9. Ibid., 39.

10. Charles Lee, "Environmental Justice: Building a Unified Vision of Health and the Environment," *Environmental Health Perspectives* 110, Suppl. 2 (2002): 141.

11. Ibid., 142.

12. Rachel Morello-Frosch, Manuel Pastor Jr., Carlos Porras, and James Sadd, "Environmental Justice and Regional Inequality in Southern California," *Environmental Health Perspectives* 110, Suppl. 2 (2002): 153.

13. Lee, "Environmental Justice,"142.

14. Ibid., 143.

15. Faber and Krieg, "Unequal Exposure to Ecological Hazards," 287.

16. Ibid.

17. Morello-Frosch et al., "Environmental Justice and Regional Inequality in Southern California," 153.

18. Ibid.

19. The "big ten" include: Defenders of Wildlife, Environmental Defense Fund, Greenpeace, National Audubon Society, National Wildlife Federation, Natural Resources Defense Council, Nature Conservancy, Sierra Club, Wilderness Society, World Wildlife Fund.

# Index

Neighborhood Citizens of Northwest, grassroots environmental justice organization, Ocala, FL, 23

Neighbors for Clean Air, grassroots environmental justice organization, Marietta, OH, 139, 150

Nelson, Mary, CEO, Bethel New Life Inc., 303–304

Newtown Creek Alliance, grassroots environmental justice organization, 252

Nickerson, Gwendolyn, resident, Corpus Christi, 106–107

Norcutt, Devon, resident, Fallon, NV, 267–268

Norcutt, Vonda, resident, Fallon, NV, 267–268

Ocala, FL, 19–40

Ohio Citizen Action (OCA), 121, 129

Oozeva, Conrad, resident, St. Lawrence Island, AK, 237–238

Organizing techniques, 9–11
citizen air monitoring, 32, 35,
city hall protests, 25, 32–33
community outreach 82–85
demonstrations, 174, 190–191
health surveys, conducted by local residents, 12, 84, 162–163, 210–212, 308–309
lawsuits (*see* Lawsuits)
letter writing campaigns, 149
media work, 10, 20, 100, 309
national experts, 10–11
research, by grassroots activists, 11, 82, 100–101, 161, 198
symptom logs, 10

Ornelia, Mary Lou, resident, San Antonio, TX, 188

Owens, Jean, resident, Addyston, OH, 122

Pace, David, reporter, Associated Press, 5

Pacific Gas & Electric (PG&E), Daly City, CA, 195
emissions, 201–202, 205–209

Paul, Laura, resident, Port Arthur, TX, 84

Penayah, Harriet, resident, St. Lawrence Island, AK, 244

Pensacola, FL, 41–69

People for Children's Environmental Health and Justice, grassroots environmental justice organization, Daly City, CA, 205

Perica, Ken, director, health and safety, Lanxess, 123

Perozzi, Sebastian, resident, Greenpoint, NY, 258

Phelan, Carinsa, resident, Fallon, NV, 273–274

Pinkney, Brenda, resident, Tallevast, FL, 170

Port Arthur, TX, 73–97

Pryor, Lewis, resident, Tallevast, FL, 166

Public housing near high-emission facilities, 74, 91, 215–216

Pungowiyi, Delbert, grassroots activist leader, St. Lawrence Island, AK, 222, 224

Pungowiyi, Perry, resident, St. Lawrence Island, AK, 224–225

Rau, Rebecca, resident, Fallon, NV, 269

Reed, Leroy, resident/activist, Ocala, FL, 19

Reed, Ruth, grassroots activist leader, Ocala, FL, 1, 19, 31, 34–35, 37

Relocation of residents exposed to industrial/military contaminants
ethics of, 68
Love Canal, NY, 52
Pensacola, FL, 52
racial bias in compensation for, 54–55
Times Beach, MO., 52

Richie, Jay, vice president, Lanxess, 123

Riverkeeper, 247, 252–253, 255

Rivers, Sarenyah, resident, Fallon, NV, 273–274

Printed in the United States
by Baker & Taylor Publisher Services